Lecture Notes in Computer Science 2066

Edited by G. Goos, J. Hartmanis and J. van Leeuwen

Springer
Berlin
Heidelberg
New York
Barcelona
Hong Kong
London
Milan
Paris
Singapore
Tokyo

Olivier Gascuel   Marie-France Sagot (Eds.)

Computational Biology

First International Conference on Biology,
Informatics, and Mathematics, JOBIM 2000
Montpellier, France, May 3–5, 2000
Selected Papers

Springer

Series Editors

Gerhard Goos, Karlsruhe University, Germany
Juris Hartmanis, Cornell University, NY, USA
Jan van Leeuwen, Utrecht University, The Netherlands

Volume Editors

Olivier Gascuel
Laboratoire d'Informatique, de Robotique et de Microelectronique de Montpellier
161 rue Ada, 34392 Montpellier Cedex 5, France
E-mail: gascuel@lirmm.fr

Marie-France Sagot
Institut Pasteur, Laboratoire d'Algorithmique Combinatoire
28, rue du Dr. Roux, 75724 Paris Cedex 15, France
E-mail: sagot@pasteur.fr

Cataloging-in-Publication Data applied for

Die Deutsche Bibliothek - CIP-Einheitsaufnahme

Computational biology : selected papers / First International Conference on
Biology, Informatics, and Mathematics, JOBIM 2000, Montpellier, France,
May 3 - 5, 2000. Olivier Gascuel ; Marie-France Sagot (ed.). - Berlin ;
Heidelberg ; New York ; Barcelona ; Hong Kong ; London ; Milan ; Paris ;
Singapore ; Tokyo : Springer, 2001
    (Lecture notes in computer science ; Vol. 2066)
    ISBN 3-540-42242-0

CR Subject Classification (1998): F.2.2, E.4, E.1, G.2, J.3

ISSN 0302-9743
ISBN 3-540-42242-0 Springer-Verlag Berlin Heidelberg New York

Springer-Verlag Berlin Heidelberg New York
a member of BertelsmannSpringer Science+Business Media GmbH

http://www.springer.de

© Springer-Verlag Berlin Heidelberg 2001
Printed in Germany

Typesetting: Camera-ready by author, data conversion by PTP-Berlin, Stefan Sossna
Printed on acid-free paper   SPIN: 10781721   06/3142   5 4 3 2 1 0

Preface

The papers contained in this volume were presented at the first annual JO-BIM conference ("Journées Ouvertes: Biologie, Informatique et Mathématique" – Open Days: Biology, Informatics and Mathematics), held May 3–5, 2000, at the ENSA-M ("École Nationale Supérieure d'Agronomie de Montpellier"), Montpellier, France. The beautiful park-like ENSA-M campus contains trees that are several hundred years old, and is only a ten minute walk from the center of the old town.

The idea of JOBIM was born two years ago, at the Place Bellecour in Lyon. We were strolling towards one of Lyon's fine restaurants. It was late one evening in April, during the RECOMB'98 conference. It seemed important to us to organize a Francophone conference that would be open to all people working at the frontier between biology, computer science, mathematics, and physics, and interested in the analysis, comparison, and exploitation of genomic data. JOBIM covers word algorithms, comparative genomics, evolution and phylogeny, data and knowledge bases, functional analysis, genome annotation, graph theory and combinatorics applied to biology, macromolecular structures (RNA and proteins), metabolic pathways and regulatory networks, statistics and classification.

We wanted JOBIM to be an annual meeting which would take place in surroundings facilitating exchanges between people, and where younger researchers in particular would be encouraged to present their work. The idea was greeted very favourably, and a program committee was rapidly set up. The initiative was well timed because at the start of the summer of 1999 the French Ministry of Research, through the Genome program, had decided to give new incentives to work in this interdisciplinary area. The IMPG project (Informatics, Mathematics, and Physics for Genomics) was created as a consequence of that decision.

The IMPG project provided an important part of the funding for JOBIM, established a precise scientific context for the conference, and assisted with publicity. Following the call for papers, we received 67 contributions. We chose to select as many papers as possible. There were 60 presentations in total, 20 long presentations for works of general interest and 40 short presentations for specialized contributions. This approach enabled a large number of researchers to present and discuss their results. There were also five invited talks, from Philipp Bucher, Christian Gautier, Gene Myers, David Sankoff, and Mike Steel. The presence of such guests contributed greatly to the scientific interest of the conference and all are gratefully thanked.

A second much stricter selection was then made for the papers that would appear in an English version of the proceedings. These are the 13 papers the reader will find in this volume. The subjects range over as broad an area as the 60 originally selected.

JOBIM was an unexpectedly huge success, with around 330 participants, showing that the field is rapidly maturing in France and the other French-speaking countries. This includes Belgium, Canada, and Switzerland. The next JOBIM will take place in Toulouse, in the south of France, May 30 to June 1, 2001. It is hoped that JOBIM 2001 will attract an even greater number of participants than the first JOBIM.

JOBIM 2000 was organized by the LIRMM at the University of Montpellier II. The Local Organizing Committee of JOBIM 2000 was headed by Gilles Caraux. The efforts of all, and general support of the French computational biology and bioinformatics community, are gratefully acknowledged.

March 2001   Olivier Gascuel
                                                       Marie-France Sagot
                                                          Program Chairs

Program Committee

Joël Alexandre
Philipp Bucher
Maxime Crochemore
Pierre Darlu
Laurent Duret
Patrick Forterre
Olivier Gascuel
Jean-François Gibrat
Alain Guénoche
Bernard Jacq
Hans-Peter Lenhof
François Major
Hervé Philippe
François Rechenmann
Éric Rivals
Claude Thermes
Alain Viari
Éric Westhof

Vincent Berry
Dominique Cellier
Antoine Danchin
Gilbert Deléage
Gwenaëlle Fichant
Nicolas Galtier
Christine Gaspin
Manolo Gouy
Jacques Haiech
Richard Lavery
Arthur Lesk
Eugene W. Myers
Bernard Prum
Jean-Loup Risler
Pierre Rouzé
Marie-France Sagot
Tandy Warnow

Organizing Committee

Gilles Caraux – OC Chair
Vincent Berry
Éric Rivals
Philippe Vismara

Isabelle Mougenot
Véronique Sals-Vettorel
Corine Ziegler

Additional referees for the LNCS volume

Andreas Dress
Alain Jean-Marie
Bruno Leclerc
Jacques Nicolas
Mike Steel

Roderic Guigó
Ned Lamb
Vincent Moulton
Cecilia Saccone
Denis Thieffry

Table of Contents

Speeding Up the DIALIGN Multiple Alignment Program by Using the 'Greedy Alignment of BIOlogical Sequences LIBrary' (GABIOS-LIB)

Saïd Abdeddaïm[1] and Burkhard Morgenstern[2]

[1] LIFAR - ABISS, Faculté des Sciences et Techniques, Université de Rouen,
76821 Mont-Saint-Aignan Cedex, France,
Said.Abdeddaim@dir.univ-rouen.fr
[2] AVENTIS Pharma, Rainham Road South, Essex RM10 7XS, UK.
Present address: MIPS, Max-Planck-Institut für Biochemie, Am Klopferspitz 18a,
82152 Martinsried Germany
morgenstern@mips.biochem.mpg.de

Abstract. A sensitive method for multiple sequence alignment should be able to align local motifs that are contained in some but not necessarily in all of the input sequences. In addition, it should be possible to integrate various of such *partial* local alignments into one single multiple output alignment. This leads to the question of *consistency* of partial alignments. Based on a new set-theoretical definition of sequence alignment, the consistency problem is discussed theoretically, and a recently developed library of C functions for consistency calculation (GABIOS-LIB) is described. GABIOS-LIB has been integrated into the DIALIGN alignment program to carry out consistency tests during the multiple alignment procedure. While the resulting alignments are exactly the same as with the previous version of DIALIGN, the running time of the program has been crucially improved. For large data sets, the new version of DIALIGN is up to 120 times faster than the old version.
Availability: http://bibiserv.TechFak.Uni-Bielefeld.DE/dialign/

Keywords: multiple sequence alignment, partial alignment, consistency, consistent equivalence relation, greedy lgorithm.

1   Introduction

Traditionally, there are two different approaches to sequence alignment: *global* methods that align sequences over their entire length [8,21,26,25] and *local* methods that try to align the most highly conserved sub-regions of the input sequences [24,23,3,13]. One problem with these approaches is that it is often not known in advance if sequences are globally or only locally related. A versatile alignment tool should align those regions of the input sequences that are sufficiently similar to each other but it would *not* try to align the non-related parts of the sequences. Thus, such a program would return a global alignment whenever sequences are globally related but a local alignment if only local homology can be

O. Gascuel, M.-F. Sagot (Eds.): JOBIM 2000, LNCS 2066, pp. 1–11, 2001.
© Springer-Verlag Berlin Heidelberg 2001

detected. One possible way to achieve this is to integrate statistically significant (partial) local alignments $P_1, \ldots, P_k$ into one resulting output alignment A.

The idea to generate sequence alignments by combining partial alignments of local similarities is not new. Various authors have proposed to generate *pairwise* local or global alignments by *chaining fragment alignments*, see Wilbur and Lipman [32], Eppstein *et al.* [7], and Chao and Miller [4]. These authors have developed time and space efficient fragment-chaining algorithms for near-optimal alignment in the sense of the traditional Needleman-Wunsch [21] and Smith-Waterman [24] objective functions. Joseph *et al.* [11] have proposed a greedy algorithm that is based on statistically significant segment pairs.

Algorithms that integrate local alignments have also been proposed for multiple alignment. Here, the problem is to decide whether a collection of local alignments is *consistent*. Informally, we call a set $\{P_1, \ldots, P_k\}$ of partial alignments consistent if an alignment A of the input sequences exists such that every pair of residues that is aligned by one or several of the alignments P_i is also aligned by A. A formal definition of our notion of consistency will be given in the next section. The question of consistency is easy to decide if each local alignment P_i involves *all* of the input sequences. Vingron and Argos [30], Vingron and Pevzner [31] and Depiereux *et al.* [6,5] have proposed multiple alignment methods that search for motifs that are simultaneously contained in all input sequences. In this case, a sufficient condition for consistency is that, for any two local alignments P_i and P_j, either $P_i \ll P_j$ or $P_j \ll P_i$ holds where $P_i \ll P_j$ means that in every sequence, residues aligned by P_i are to the left of residues aligned by P_j. From a biological point of view, however, it is desirable to allow for homologies involving not all but only some of the input sequences. A multiple alignment program that finds only those similarities that are present in *all* sequences in a given input data set will necessarily miss many biologically important homologies.

Recently, we have introduced three heuristics for multiple alignment that integrate partial local alignments not necessarily involving all of the input sequences. These methods generate multiple alignments in a greedy way by incorporating local partial alignments one-by-one into a resulting multiple alignment. SOUNDALIGN [1] assembles multiple alignments from *blocks* of un-gapped local alignments that may involve two or more sequences, DIALIGN [15,17,18] uses un-gapped segment pairs – so-called *fragments* or *diagonals* –, and TWOALIGN [2] combines pairwise local alignments in the sense of Smith and Waterman [24] to obtain a final multiple alignment. During the greedy procedures, these three programs have to test new partial alignments for *consistency* with those alignments that have already been accepted for the final alignment. To this end, they store and update certain data structures that are called *transitivity frontiers* for SOUNDALIGN and TWOALIGN and *consistency bounds* for DIALIGN.

DIALIGN has been successfully used to detect local homologies in nucleic acid and protein sequences. In a recent study, Göttgens *et al.* [10] have used DIALIGN to align large genomic sequences from human, mouse and chicken. In the human/mouse alignment, multiple peaks of homology were found, some of which precisely correspond to known enhancers. In addition, the DIALIGN multi-alignment of human, mouse and chicken revealed a new region of homology that was then experimentally shown to be a previously unknown enhancer, see also Miller [14] for a discussion of these results. Thompson *et al.* [28], have used the BAliBASE of benchmark alignments [27] to systematically compare the most widely used programs for multiple protein alignment. Here, DIALIGN was reported to be the most successful *local* method. It also performed well on *globally* related sequence sets though here CLUSTAL W [26], PRRP [9] and SAGA [22] were superior. The paper by Thompson *et al.*, however, addressed also a major weakness of DIALIGN: it is considerably slower than progressive alignment methods. Aligning a set of 89 histone sequences took as much as 13,649 *s* with DIALIGN compared to 161 *s* with CLUSTAL. This may not be a serious problem if only a single protein family is studied. However, with the huge amount of genomic sequence data that are now available, automatic alignment of whole data bases of sequence families has become a routine task, see, for example, [12,29]. Here, program running time is a crucial criterion in choosing a suitable alignment method.

Test runs have shown that for large sequence sets, the procedure of updating the consistency bounds was by far the most time-consuming step in previous versions of DIALIGN. Abdeddaïm has recently developed a library of C functions called *GABIOS-LIB* (Greedy Alignment of BIOlogical Sequences LIBrary) that can be used to efficiently calculate the transitivity frontiers and to consistency-check (partial) local alignments that may have been produced by arbitrary methods. We have integrated GABIOS-LIB into DIALIGN to speed up the consistency check for fragments (segment pairs). In the present paper, the time and space complexity of GABIOS-LIB is analysed theoretically and compared to the method that was previously used in DIALIGN. Experiments with both artificial and real sequence data demonstrate that GABIOS-LIB is far more efficient than the previous procedure. In our test examples, the new version 2.1 of DIALIGN is up to 120 times faster than version 2.0 while the resulting alignments are exactly the same. In addition, GABIOS-LIB has reduced the amount of computer memory used by DIALIGN.

2   The Consistency Problem for Partial Alignments

2.1   Definitions and Notations

Let $\mathcal{S} = \{S_1, \ldots, S_N\}$ be a sequence family and let X be the set of all *sites* of $\mathcal{S}$ where a site $x = [i, p]$ represents the p-th position in the i-th sequence. On X, we define a partial order relation $\preceq$ such that for any two sites $x = [i, p]$ and $x' = [i', p']$, $x \preceq x'$ holds if and only if both $i = i'$ and $p \leq p'$ are true. In the language of order theory, $\preceq$ is the *direct sum* of the 'natural' linear order relations that are given on the individual sequences. Every binary relation R on X *extends* the relation $\preceq$ to a *quasi order relation* $\preceq_R = (\preceq \cup R)_t$ on X, where S_t denotes the transitive closure of a relation S, i.e. the smallest transitive relation containing S, see Fig. 1.

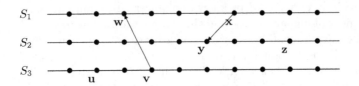

Fig. 1. A relation $R = \{(v,w),(x,y)\}$ (represented by arrows) defined on the set X of all *sites* (black dots) of a sequence family $\mathcal{S} = \{S_1, S_2, S_3\}$ *extends* the *partial* order relation $\preceq$ on X to a *quasi partial* ordering $\preceq_R = (\preceq \cup R)_t$. We have, for example, $u \preceq v$, vRw, $w \preceq x$, xRy, $y \preceq z$ and therefore $u \preceq_R z$.

We call a relation R on X *consistent* if the extended relation $\preceq_R$ *preserves* the linear order relations on the individual sequences, formally: if all restrictions of $\preceq_R$ to the individual sequences *coincide* with their respective 'natural' linear order relations. In other words, the requirement is that for any two sites x and y that belong to the same sequence, $x \preceq_R y$ implies $x \preceq y$. Moreover, we call a set $\{R_1, \ldots, R_n\}$ of relations consistent if the union $\cup_i R_i$ is consistent, we say that R_1 is consistent with R_2 if $\{R_1, R_2\}$ is consistent, and a pair $(x,y) \in X^2$ is called consistent with a relation R if $\{(x,y)\}$ is consistent with R.

As proposed in [17], a (partial) *alignment* A of the family $\mathcal{S}$ can be defined as a *consistent equivalence relation* on the set X where we write $(x,y) \in A$ or xAy if the sites x and y are either aligned by A or identical. As an equivalence relation, an alignment A partitions X into equivalence classes $[x]_A = \{y \in X : (x,y) \in A\}$. It can be shown that an equivalence relation A on X is an alignment in the sense of the above definition if and only if it is possible to introduce *gap characters* into the sequences S_i such that the equivalence classes $[x]_A, x \in X$ are precisely those sets of sites that are in the same column of the resulting two-dimensional array, see [20] for more details.

A common feature of greedy multiple alignment algorithms is that they include partial alignments $P_1, \ldots, P_k$ one after the other into a growing multiple alignment – always provided that a new alignment P_i is *consistent* with those alignments that have been included previously. Formally, a monotonously increasing set $A_1 \subset \ldots \subset A_k$ of alignments is defined by

$$A_1 = P_1$$
$$A_i = \begin{cases} (A_{i-1} \cup P_i)_e & \text{if } P_i \text{ is consistent with } A_{i-1} \\ A_{i-1} & \text{otherwise,} \end{cases} \quad i = 2, \ldots, k. \qquad (1)$$

A final alignment A is then obtained as the largest alignment $A = A_k$ of this set. Therefore, every greedy alignment approach has to resolve the question of consistency: at any stage of the alignment procedure, it must be known which pairs $(x,y) \in X^2$ are still *alignable* without leading to inconsistencies with the current alignment A_i, i.e. with those pairs of sites that have already been accepted for the final alignment.

2.2 Transitivity Frontiers and Consistency Bounds

An alignment A of a sequence family $\mathcal{S}$ imposes for every site $x \in X$ and every sequence $S_i \in \mathcal{S}$ a lower bound $\underline{b}_A(x, i)$ and an upper bound $\overline{b}_A(x, i)$ such that a site $y = [i, p] \in S_i$ is alignable with x without leading to inconsistencies with A if and only if $\underline{b}_A(x, i) \le p \le \overline{b}_A(x, i)$ holds, see Fig. 2 for an example. Formally, we define

$$\underline{b}_A(x, i) = \min\{p : (x, [i, p]) \text{ consistent with } A\}$$

and

$$\overline{b}_A(x, i) = \max\{p : (x, [i, p]) \text{ consistent with } A\}.$$

In order to test fragments for consistency during the greedy procedure, previous versions of DIALIGN calculated and updated these *consistency bounds*.

GABIOS-LIB is using a so-called *transitivity frontiers* to carry out the same consistency check. Here, the *predecessor frontier* $Pred_A(x, i)$ is defined as the position of the right-most site y in sequence S_i such that $y \preceq_A x$ is true, and the *successor frontier* $Succ_A(x, i)$ is defined accordingly as the position of the left-most site y in sequence S_i with $x \preceq_A y$ so we have

$$Pred_A(x, i) = \max\{p : [i, p] \preceq_A x\}$$

and

$$Succ_A(x, i) = \min\{p : x \preceq_A [i, p]\}.$$

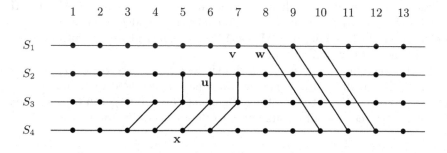

Fig. 2. For an alignment A (bold lines) and a site x, the transitivity frontiers with respect to a sequence S_i *coincide* with the corresponding consistency bounds if x is aligned to some site in S_i. For example, site u in S_2 is aligned with x, so we have $\overline{b}_A(x, 2) = Succ_A(x, 2) = 6$. In sequence S_1, on the other hand, v is the right-most site that can be aligned with x, but w is the left-most site with $x \preceq_A w$, so we have $\overline{b}_A(x, 1) = 7$ but $Succ_A(x, 1) = 8$.

The transitivity frontiers are related to the consistency bounds in the following way: if x is already aligned to some site $[i, p]$ in sequence S_i, then the predecessor and successor frontiers of x with respect to S_i both equal p and they *coincide* with the consistency bounds, i.e. one has

$$Pred_A(x, i) = Succ_A(x, i) = \underline{b}_A(x, i) = \overline{b}_A(x, i) = p.$$

This is, for example, the case for the frontiers and bounds of x with respect to S_2 in Fig. 2. In contrast, if no site in S_i is aligned with x, one has

$$Pred_A(x, i) = \underline{b}_A(x, i) - 1$$

and

$$Succ_A(x, i) = \overline{b}_A(x, i) + 1$$

as is the case with the corresponding frontiers and bounds with respect to S_1 in Fig. 2. Therefore, if it is known for every site x and every sequence S_i whether x is aligned with some site in S_i, transitivity frontiers are easily obtained from the consistency bounds and vice versa, so both data structures are equivalent in that they contain the same information about which residue pairs are alignable under the consistency constraints imposed by a given alignment A.

3   GABIOS-LIB

The Greedy Alignment of BIOlogical Sequences LIBrary (GABIOS-LIB) is a set of functions implemented in ANSI C by Abdeddaïm. These functions can be used by any greedy alignment program in order to test in constant time which sites in a sequence S_j are alignable with a site x of an other sequence S_i. Each time two sites are aligned during the greedy procedure, GABIOS-LIB updates the transitivity frontiers using the incremental algorithm EdgeAddition presented in [2]. In addition, GABIOS-LIB uses some ideas first introduced in [1] to further reduce computing time and memory.

In this section, we discuss how the successor frontiers $Succ_A(x, i)$ for an alignment A are affected if a (partial) alignment P is added to A and how the frontiers $Succ_B(x, i)$ for the resulting alignment $B = (A \cup P)_e$ can be calculated. For symmetry reasons, all results apply to the predecessor frontiers as well.

3.1   The Incremental Algorithm EdgeAddition

First of all, a simple but important observation is that if a new partial alignment P is added to an existing alignment A, the frontiers with respect to the new alignment $B = (A \cup P)_e$ need to be calculated only if B is actually different from A which is the case if and only if P is not already a subset of A. EdgeAddition stores for every site x and every sequence S_j the information whether x is already aligned with some residue from sequence S_j; this information is used to check if a new alignment P is already contained in A.

If and only if $P \not\subseteq A$ holds, the transitivity frontiers $Succ_B$ are different from the frontiers $Succ_A$. In general, however, the frontiers will change not for all but only for some sites, and the computing time can be minimized by identifying those sites. For simplicity, we consider the simplest case where a single pair of sites (x, y) is added to A.

Observation 1 *Let A be an alignment of a sequence family S and (x, y) a pair of sites that is consistent with A. Let $B = (A \cup \{(x, y)\})_e$ be the alignment that is obtained by 'adding' (x, y) to A. Then for every site u in a sequence S_i and for every sequence S_j the successor frontiers of B are*

$$Succ_B(u, j) = \begin{cases} \min\{Succ_A(u, j), Succ_A(y, j)\} & \text{if } u \preceq_A x \\ \min\{Succ_A(u, j), Succ_A(x, j)\} & \text{if } u \preceq_A y \\ Succ_A(u, j) & \text{otherwise.} \end{cases}$$

It follows that the transitivity frontiers can change only for those sites u for which either $u \preceq_A x$ or $u \preceq_A y$ is true.

3.2   Further Reduction of Computing Time and Memory

In order to further reduce the computational costs for calculating the consistency frontiers, GABIOS-LIB uses the following two facts. (1) If two sites x and y are aligned by A, i.e. if xAy holds, then they have necessarily the same frontiers $Succ_A(x, i) = Succ_A(y, i)$ for $i = 1, \ldots, N$. Therefore, rather than processing the transitivity frontiers for all *individual* sites x, GABIOS-LIB stores and updates the frontiers for those *equivalence classes* $[x]_A$ that consist of more than one single site. (2) Let $x = [i, p]$ be an *orphan* site in the i-th sequence, i.e. a site that is *not* aligned with any other site $y \neq x$. Then the successor frontiers $Succ_A(x, j)$ with respect to all sequences $S_j \neq S_i$, *coincide* with the corresponding frontiers of the left-most non-orphan site $y = [i, p']$ with $p' > p$. In Fig. 2, for example, $v = [1, 7]$ is an orphan site in S_1. The left-most non-orphan site $[1, p']$ with $p' > 7$, is the site $w = [1, 8]$. Therefore for $j \neq 1$, the successor frontiers $Succ_A(v, j)$ coincide with the corresponding frontiers for w, i.e. we have $Succ_A(v, j) = Succ_A(w, j)$ for $j = 2, 3, 4$. Thus, instead of storing the transitivity frontiers for an orphan site x, the corresponding site y can be stored in a tabular nextClass by defining nextClass$[x] = p'$. This way, the frontiers $Succ_A(x, j)$ of an orphan site x can be established in constant time each time a new pair of sites is aligned.

4   Time and Space Efficient Multiple Segment-by-Segment Alignment

In order to construct a multiple alignment of a sequence family S, DIALIGN calculates in a first step optimal pairwise alignments for all possible pairs of input sequences as explained in [16]. Optimality, in this context, refers to the segment-based objective function used in DIALIGN as defined in [19,15], i.e. an optimal pairwise alignment is a *chain* of fragments (gap-free segment pairs)

Table 1. Running time t_0 and t_1 of versions 2.0 and 2.1 of DIALIGN. Version 2.0 is using the old method of calculating the consistency bounds while version 2.1 is calculating the transitivity frontiers using GABIOS-LIB. Sequences are (I) *independent* and (II) *identical* random sequences of length 100, (III) Immunoglobulin domain sequences of average length 63.3 and 20% average identity, and (IV) Ribosomal protein L7/L12 C-terminal domain sequences of average length 119.3 and 53% average identity. The running time improvement achieved by GABIOS-LIB strongly increases with the number N of sequences to be aligned. For the two sets of real-world sequences, figures are comparable to the case of identical random sequences where an improvement of up to 120 was achieved.

		DIALIGN 2.0			DIALIGN 2.1			2.0/2.1
	N	t_0	t_0/N^3	t_0/N^4	t_1	t_1/N^2	t_1/N^3	t_0/t_1
(I)	25	22	$1.4\ 10^{-3}$	$5.6\ 10^{-5}$	10	0.016	$6.4\ 10^{-4}$	2.2
	50	168	$1.3\ 10^{-3}$	$2.6\ 10^{-5}$	41	0.016	$3.3\ 10^{-4}$	4.1
	75	601	$1.4\ 10^{-3}$	$1.8\ 10^{-5}$	100	0.017	$2.3\ 10^{-4}$	6.0
	100	1402	$1.4\ 10^{-3}$	$1.4\ 10^{-5}$	220	0.022	$2.2\ 10^{-4}$	6.8
	150	5725	$1.6\ 10^{-3}$	$1.1\ 10^{-5}$	576	0.025	$1.7\ 10^{-4}$	10.0
	200	14353	$1.7\ 10^{-3}$	$8.9\ 10^{-6}$	1874	0.046	$2.3\ 10^{-4}$	7.6
(II)	25	46	$2.9\ 10^{-3}$	$1.1\ 10^{-4}$	10	0.016	$6.5\ 10^{-4}$	4.6
	50	640	$5.1\ 10^{-3}$	$1.0\ 10^{-4}$	43	0.017	$3.4\ 10^{-4}$	14.9
	75	3282	$7.7\ 10^{-3}$	$1.0\ 10^{-4}$	104	0.018	$2.4\ 10^{-4}$	31.6
	100	10557	$1.0\ 10^{-2}$	$1.0\ 10^{-4}$	200	0.020	$2.0\ 10^{-4}$	52.8
	150	52431	$1.5\ 10^{-2}$	$1.0\ 10^{-4}$	555	0.024	$1.6\ 10^{-4}$	94.5
	200	177423	$2.2\ 10^{-2}$	$1.1\ 10^{-4}$	1429	0.035	$1.7\ 10^{-4}$	124.2
(III)	25	25	$1.6\ 10^{-3}$	$6.4\ 10^{-5}$	7	0.011	$4.4\ 10^{-4}$	3.5
	50	341	$2.7\ 10^{-3}$	$5.4\ 10^{-5}$	29	0.011	$2.3\ 10^{-4}$	11.7
	75	1597	$3.7\ 10^{-3}$	$5.0\ 10^{-5}$	73	0.012	$1.7\ 10^{-4}$	21.8
	100	5046	$5.0\ 10^{-3}$	$5.0\ 10^{-5}$	147	0.014	$1.4\ 10^{-4}$	34.3
(IV)	25	61	$3.9\ 10^{-3}$	$1.5\ 10^{-4}$	14	0.022	$8.9\ 10^{-4}$	4.3
	50	844	$6.7\ 10^{-3}$	$1.3\ 10^{-4}$	63	0.025	$5.0\ 10^{-4}$	13.3
	75	4157	$9.8\ 10^{-3}$	$1.3\ 10^{-4}$	149	0.026	$3.5\ 10^{-4}$	27.8
	100	11619	$1.1\ 10^{-2}$	$1.1\ 10^{-4}$	288	0.028	$2.8\ 10^{-4}$	40.3

$f_1 \ll \ldots \ll f_k$ such that $\sum_{i=1}^{k} w(f_i)$ is maximal. Here, w is a weighting function defined on the set of all possible fragments, and $f_i \ll f_j$ means that the end positions of f_i are both strictly smaller than the respective beginning positions of f_j. Fragments from the respective optimal pairwise alignments are then greedily integrated into a resulting multiple alignment A. Let L be the maximum length of the sequences $S_1, \ldots, S_N$. Since $O(N^2)$ pairwise alignments are performed each of which consisting of at most L fragments, $O(N^2 L)$ fragments are to be checked for consistency. Updating the consistency bounds takes $O(N^2)$ time if a new fragment is accepted for alignment. Since previous versions of DIALIGN calculated the consistency bounds for *every* fragment that was accepted for alignment, the

Table 2. Number of integers allocated by DIALIGN 2.0 and DIALIGN 2.1 for storing consistency bounds (#CB) and transitivity frontiers (#TF), respectively. Sequence sets are as in Table 1.

N		$\#CB$	$\#TF$	$\frac{\#CB}{\#TF}$		$\#CB$	$\#TF$	$\frac{\#CB}{\#TF}$
25		125,000	5,000	25.0		85,700	11,500	7.5
50	(I)	500,000	10,000	50.0	(III)	346,600	43,900	7.9
75		1,125,000	15,000	75.0		780,000	76,650	10.2
100		2,000,000	20,000	100.0		1,382,800	142,400	9.7
25		125,000	28,050	4.5		148,800	13,300	11.2
50	(II)	500,000	106,800	4.7	(IV)	609,600	75,600	8.1
75		1,125,000	204,150	5.5		1,356,450	145,350	9.3
100		2,000,000	375,000	5.3		2,403,800	212,200	11.3

worst-case time complexity of the entire greedy procedure was $O(N^4 L)$. Table 1 shows that the running time of DIALIGN 2.0 is in fact proportional to N^4 if *identical* sequences are aligned where *all* fragments from the optimal pairwise alignments are necessarily consistent and are therefore integrated into A.

Let us consider a consistent set of pairs $\{(x_1, y_1), \ldots, (x_m, y_m)\}$ that are successively integrated into a growing set of alignments $A_1 \subset \ldots \subset A_m$ by defining $A_i = \{(x_1, y_1), \ldots, (x_i, y_i)\}_e$. It was shown in [2] that if each pair (x_i, y_i) is actually *new*, i.e. not yet contained in the previous alignment A_{i-1}, then GABIOS-LIB takes $O(N^2 m + |X|^2)$ time to update the transitivity frontiers during the procedure of integrating all m pairs. It was also explained in [2] that there can be at most $|X|$ 'new' pairs of sites – every additional pair would either be inconsistent or would already be contained in the current alignment. Therefore, GABIOS-LIB takes $O(N^2|X| + |X|^2) = O(N^3 L + N^2 L^2)$ time to compute the transitivity frontiers while integrating an *arbitrary* set of pairs. Note that this complexity analysis is a worst-case estimate. As shown in Table 1, the real running time of GABIOS-LIB is better than $O(N^3)$. The new version 2.1 of DIALIGN is still slower than the most widely used multi-alignment program CLUSTAL W [26], but the difference in running time is now reduced to a factor of about 10. Parameter optimization should further decrease the running time of DIALIGN.

In the previous version of DIALIGN, the consistency bounds were stored for *every* site $x \in X$. Since for every x, $2N$ integer values had to be considered – upper and lower bounds with respect to all N sequences –, computer memory had to be allocated for exactly $2N|X|$ integer values. By contrast, GABIOS-LIB stores $2N$ transitivity frontiers only for *non-orphan* equivalence classes. For orphan sites, single integers are stored that refer to the next non-orphan site in the same sequence. In the *worst case*, the number of non-orphan equivalence classes is in the order of $|X|$, so the space complexity for GABIOS-LIB is *upper-*

bounded by $O(|X|N) = O(N^2L)$ which was also the *real* space complexity of the old version of DIALIGN. Table 2 shows, however, that the actual memory requirement for storing the transitivity frontiers with GABIOS-LIB is far smaller than for storing the consistency bounds in DIALING 2.0.

References

1. S. Abdeddaïm. Fast and sound two-step algorithms for multiple alignment of nucleic sequences. In *Proceedings of the IEEE International Joint Symposia on Intelligence and Systems*, pages 4–11, 1996.
2. S. Abdeddaïm. Incremental computation of transitive closure and greedy alignment. In *Proc. of 8-th Annual Symposium on Combinatorial Pattern Matching*, volume 1264 of *Lecture Notes in Computer Science*, pages 167–179, 1997.
3. S. F. Altschul, W. Gish, W. Miller, E. M. Myers, and D. J. Lipman. Basic local alignment search tool. *J. Mol. Biol.*, 215:403–410, 1990.
4. K.-M. Chao and W. Miller. Linear-space algorithms that build local alignments from fragments. *Algorithmica*, 13:106–134, 1995.
5. E. Depiereux, G. Baudoux, P. Briffeuil, I. Reginster, X. D. Boll, C. Vinals, and E. Feytmans. Match-Box server: a multiple sequence alignment tool placing emphasis on reliability. *CABIOS*, 13:249–256, 1997.
6. E. Depiereux and E. Feytmans. Match-box: a fundamentally new algorithm for the simultaneous alignment of several protein sequences. *CABIOS*, 8:501–509, 1992.
7. D. Eppstein, Z. Galil, R. Giancarlo, and G. Italiano. Sparse dynamic programming I: Linear cost functions. *J. Assoc. Comput. Mach.*, 39:519–545, 1992.
8. O. Gotoh. An improved algorithm for matching biological sequences. *J. Mol. Biol.*, 162:705–708, 1982.
9. O. Gotoh. Significant improvement in accuracy of multiple protein sequence alignments by iterative refinement as assessed by reference to structural alignments. *J. Mol. Biol.*, 264:823–838, 1996.
10. B. Göttgens, L. Barton, J. Gilbert, A. Bench, M. Sanchez, S. Bahn, S. Mistry, D. Grafham, A. McMurray, M. Vaudin, E. Amaya, D. Bentley, and A. Green. Analysis of vertebrate scl loci identifies conserved enhancers. *Nature Biotechnology*, 18:181–186, 2000.
11. D. Joseph, J. Meidanis, and P. Tiwari. Determining DNA sequence similarity using maximum independent set algorithms for interval graphs. *Lecture Notes in Computer Science*, 621:326–337, 1992.
12. A. Krause, P. Nicodème, E. Bornberg-Bauer, M. Rehmsmeier, and M. Vingron. Www access to the systers protein sequence cluster set. *Bioinformatics*, 15:262–263, 1999.
13. C. E. Lawrence, S. F. Altschul, M. S. Boguski, J. S. Liu, A. F. Neuwald, and J. C. Wootton. Detecting subtle sequence signals: a gibbs sampling strategy for multiple alignment. *Science*, 262(5131):208–14, 1993.
14. W. Miller. So many genomes, so little time. *Nature Biotechnology*, 18:148–149, 2000.
15. B. Morgenstern. DIALIGN 2: improvement of the segment-to-segment approach to multiple sequence alignment. *Bioinformatics*, 15:211–218, 1999.
16. B. Morgenstern. A space-efficient algorithm for aligning large genomic sequences. *Bioinformatics*, in press.

17. B. Morgenstern, A. W. M. Dress, and T. Werner. Multiple DNA and protein sequence alignment based on segment-to-segment comparison. *Proc. Natl. Acad. Sci. USA*, 93:12098–12103, 1996.

18. B. Morgenstern, K. Frech, A. W. M. Dress, and T. Werner. DIALIGN: finding local similarities by multiple sequence alignment. *Bioinformatics*, 14:290–294, 1998.

19. B. Morgenstern, K. Hahn, W. R. Atchley, and A. W. M. Dress. Segment-based scores for pairwise and multiple sequence alignments. In J. Glasgow, T. Littlejohn, F. Major, R. Lathrop, D. Sankoff, and C. Sensen, editors, *Proceedings of the Sixth International Conference on Intelligent Systems for Molecular Biology*, pages 115–121, Menlo Parc, CA, 1998. AAAI Press.

20. B. Morgenstern, J. Stoye, and A. W. M. Dress. Consistent equivalence relations: a set-theoretical framework for multiple sequence alignment. Materialien und Preprints 133, University of Bielefeld, 1999.

21. S. B. Needleman and C. D. Wunsch. A general method applicable to the search for similarities in the amino acid sequence of two proteins. *J. Mol. Biol.*, 48:443–453, 1970.

22. C. Notredame and D. Higgins. SAGA: sequence alignment by genetic algorithm. *Nucleic Acids Research*, 24:1515 – 1524, 1996.

23. W. R. Pearson and D. J.Lipman. Improved tools for biological sequence comparison. *Proc. Nat. Acad. Sci. USA*, 85:2444–2448, 1988.

24. T. F. Smith and M. S. Waterman. Comparison of biosequences. *Advances in Applied Mathematics*, 2:482–489, 1981.

25. J. Stoye. Multiple sequence alignment with the divide-and-conquer method. *Gene*, 211:GC45–GC56, 1998.

26. J. D. Thompson, D. G. Higgins, and T. J. Gibson. CLUSTAL W: improving the sensitivity of progressive multiple sequence alignment through sequence weighting, position-specific gap penalties and weight matrix choice. *Nucleic Acids Research*, 22:4673–4680, 1994.

27. J. D. Thompson, F. Plewniak, and O. Poch. BAliBASE: A benchmark alignment database for the evaluation of multiple sequence alignment programs. *Bioinformatics*, 15:87–88, 1999.

28. J. D. Thompson, F. Plewniak, and O. Poch. A comprehensive comparison of protein sequence alignment programs. *Nucleic Acids Research*, 27:2682–2690, 1999.

29. J. D. Thompson, F. Plewniak, J.-C. Thierry, and O. Poch. DbClustal: rapid and reliable global multiple alignments of protein sequences detected by database searches. *Nucleic Acids Research*, 28:2919–2926, 2000.

30. M. Vingron and P. Argos. Motif recognition and alignment for many sequences by comparison of dot-matrices. *J Mol Biol*, 218(1):33–43, 1991.

31. M. Vingron and P. Pevzner. Multiple sequence comparison and consistency on multipartite graphs. *Advances in Applied Mathematics*, 16:1–22, 1995.

32. J. W. Wilbur and D. J. Lipman. The context dependent comparison of biological sequences. *SIAM J. Appl. Math.*, 44:557–567, 1984.

GeMCore, a Knowledge Base Dedicated to Mapping Mammalian Genomes

Gisèle Bronner,[1,2] Bruno Spataro,[1] Christian Gautier,[1] and François Rechenmann[2]

[1] Laboratoire de Biométrie et Biologie Évolutive, UMR – CNRS 5558, Université Claude Bernard Lyon1
43, bd du 11 novembre 1918, 69622 Villeurbanne cedex, France
{bronner,bspataro,cgautier}@biomserv.univ-lyon1.fr
[2] INRIA Rhône Alpes
655 av. de l'Europe, 38330 Montbonnot St Martin, France
Francois.Rechenmann@inrialpes.fr

Abstract. Representing genomic mapping knowledge is a complex process due to the diversity of experimental approaches. The lack of obvious equivalence between different kinds of maps[1] makes it difficult to use them simultaneously or in a cooperative way. Furthermore, comparative mapping is not limited to map comparison, it also implies the comparison of the elements they are composed of. Comparison thus implies the combination of different computational objects, corresponding to different levels of information. However, from the user's point of view, systems devoted to comparative mapping should appear to process queries in a homogeneous manner, whatever the objects they imply. We propose a data model that integrates mapping data, genetical data and sequences and takes into account the specific constraints resulting from the comparison of maps between and within different species.

1   Introduction

Comparative interspecies analysis of nucleic or proteinic sequences is helpful in understanding genome evolution and function. Comparisons can be made at different levels

- sequence comparisons: codon usage, alignments, phylogeny;
- function comparisons: protein functions, functional domains;
- structure comparisons: DNA structures, protein strutures, genome structure.

The relative position of interesting genes and markers on chromosomes can be drawn on genomic maps to identify regions where candidate sequences corresponding to the markers should lie. A large number of mapping projects are providing increasing amounts of information on the localizations of biological elements. Such information, that is the position of elements on a chromosome, can be compared as sequences or biological elements. An initial analysis suggests that chromosomic structures are relatively conserved in vertebrates [1], allowing prediction of gene positions among

---

[1] Definitions for map, markers and other biological concepts can be found in the Appendix.

O. Gascuel, M.-F. Sagot (Eds.): JOBIM 2000, LNCS 2066, pp. 12-23, 2001.

species. As a consequence, mutual enrichment of maps from different species greatly improves the mapping approach: reference species can be used to help in the analysis of species of interest. Comparative mapping thus emerges as a powerful tool to gather new knowledge about genomes.

Present mapping data management systems allow users to access biological data on single objects, but do not permit a global analysis of biological objects according to their genomic localization. However, the interest of comparative mapping is not limited to comparisons of gene localization. It can also help in resolving biological problems. For example, comparative mapping studying the *C. elegans* daf-18 ortholog has been used to characterize possible phenotypes associated with mutations of the human PTEN tumor suppressor gene [2].

Because of the lack of adequate tools for comparative mapping, we are developing GeMCore, a knowledge base dedicated to the management and comparison of mapping data, as well as functional and sequence data on mammals. GeMCore is part of the GeM project (for Genomic Mapping) which consists in the development of a core knowledge base (GeMCore) connected to graphical interfaces devoted to particular domains. Parallel to GeMCore, the user interface GeMME (for Genomic Mapping and Molecular Evolution) [3], is being developed. GeMME allows queries on several data types such as gene symbol, localization or expression of genes as well as queries related to sequence composition or evolution. It proposes numerical and graphical representation of data allowing visual display of synteny groups and map-to-map switching between species. GeMME contains several analytical tools devoted to molecular evolution. These tools can, for example, highlight specific spatial patterns of the human genome such as isochore [4] or autocorrelation of the silent mutation rate [5].

In the next sections, we will first present the requirements of any system dedicated to comparative mapping, then will describe the data model used in GeMCore.

2   Requirements of Mapping Data Systems

Comparative genomics involving many species requires the use of many data sources because:
- each mapping project has its own official database such as the Genome Data Base (GDB) [6] for the human genome or the Mouse Genome Database (MGD) [7] for the mouse genome;
- many databases of particular interest exist for different species (GDB, LDB [8], OMIM [9]).

Furthermore, the diversity of experimental genome mapping approaches (genetic, cytogenetic, physical RH maps...) produces different types of maps. These maps are not equivalent, and thus define different points of view of the same object. The diversity and heterogeneity of this information (heterogeneity of the data themselves but also of their storage format) is difficult to manage. Three major problems can be described:

Data inconsistency: mapping data comparisons can exhibit important contradictions among data, and sometime real errors. There are two kinds of inconsistency. Inconsistency can result from an erroneous representation of the data: in this case, the information is correct but expressed wrongly. For example, the insulin (Ins1) gene, which has been assigned to the 11p15 region on the human genome, was drawn around the 11p12-11p14 region in the GDB chromosome 11 integrated map. Inconsistency can also be intrinsic to the data. For example, the order of markers on a map can differ depending on the mapping method that is considered (some mapping methods based on probabilistic inference sometimes infer non-consensual marker order). Here, inconsistency cannot be suppressed by correcting the information. Thus, the system must propose explicit strategies of inconsistency management and integrate new user choices in these strategies.

Lack of formal links between data: for the human and mouse genomes, links between localization and sequences, genes and gene products are sometimes approximate and far from complete. For other species, the situation is even worse. There have been some attempts to automatically generate these links through semantic analysis of databank annotations. This led to the Virgil [10] or the LocusLink [11] protocols for the human genome. This strategy can associate the genetic name of a gene (and so its localization) to a genomic sequence, but cannot discriminate between different coding sequences within the same fragment. Moreover, alternative splicing which is very frequent in eukaryotes, makes it even more complex to assign a genetic name to a coding sequence.

Data redundancy: parallel gathering of similar data from different sources generates strong redundancy in banks. As an example, the human insulin gene sequence is recorded four times in Genbank.

Comparative mapping data management systems must be able to handle data diversity, data sources and constraints inherent to such information. They are not limited to gathering and integrating pre-existing data, but consist in distributing a clarified, completed and organized information set capable of reducing the difficulties described above.

This pre-processing implies the development of procedures specific to this stage of knowledge base building. This phase also requires the use of external programs such as Clustal [12] or Blast [13, 14] that should not be integrated in the knowledge base. As the data gathering protocol is designed for a specific problematic, it does not meet our needs for producing a generic knowledge base, although a specific interface devoted to the data gathering protocol has been developed. This interface handles some aspects of data consistency management, redundancy avoidance and data linkage as well reorganization of data gathered from diverse databases into a common format. Because certain procedures such as identification of homologies require manual expertise, the interface is currently only partly automated.

This procedure allows the building of an integrated dataset. It will constitute the starting point of the data gathering of the knowledge base. As the GeMCore implementation is still in progress, this dataset is currently used to test the system.

The dataset consist in cytogenetical descriptions of human and mouse chromosomes, location data for human and mouse genes, sequences and sequence analysis (such as G+C content, CDS length, alignment length, substitution rates...), expression data, functional homology data, evolutive relations and recombination rates.

In this dataset, links between genes and coding sequences are built from the NCBI LocusLink file for human and the MGD MRK_Sequence.rpt file for mouse. The relevance of all associations between a CDS (Coding DNA Sequence) and a gene symbol are evaluated. To discriminate between coding sequences associated to the

Table 1. Description of the dataset used in the evaluation of GeMCore.

data		Source	file size
locations	human	LDB[+2]	4964 genes
	mouse	MGD[3]	7203 genes
gene/sequence links	human	locuslink[4], Hovergen[5]	14172 links
	mouse	MGD, Hovergen	11040 links
sequence analysis		Hovergen, Gccompute	22883 seq.
CDS analysis		Hovergen, Gccompute	21066 seq.
homology		MGD	4044 cples of genes
orthology – substitution rates		Orthogen, JaDis[6]	3998 cples
cytogenetic maps	human	LDB	23 chr.
	mouse	MGD	20 chr.

same gene symbol, we use similarity information expressed through the family and redundancy numbers of the Hovergen database [15]. Hovergen redundancy numbers group under a common annotation redundant sequences from Genbank. Two sequences are said to be redundant if they have less than 1% divergence and at most 3 nucleotides of difference in length. The family numbers group homologous sequences (orthologous or paralogous genes resulting from duplication anterior to the divergence of vertebrates) under the same identifier.

For each locus symbol, LocusLink provides a list of accession numbers, that is a list of genomic fragments of Genbank. A method has been developed to generate a list of relevant couples of gene symbol / gene sequence (CDS) extracted from these fragments. First, all links between a gene symbol and a CDS are listed. For each symbol, if all CDS have the same redundancy number, they are assumed to be several entries of the same gene and are all associated to the locus symbol. If several redundancy numbers are associated to the same gene symbol, we search for a CDS having the locus symbol cited in the corresponding fields of Genbank annotations. If found, only the CDS with this redundancy number are associated to the locus symbol. To keep most of the LocusLink information, we introduce two "weak" links between locus symbols and CDS belonging to the genomic fragment associated in LocusLink:

- "alternative" CDS have a different redundancy number than the one associated to the linked CDS but belong to the same family
- "putative" CDS correspond to cases where all CDS belong to the same family, but have different redundancy number and no locus symbol in Genbank annotations.

---

[2] Location DataBase - http://cedar.genetics.soton.ac.uk/public_html/ldb.html

[3] Mouse Genome Database - http://www.informatics.jax.org

[4] LocusLink - http://ncbi.nlm.nih.gov/LocusLink/index.html

[5] Homologous Vertebrate Gene Database - http://pbil.univ-lyon1.fr/hovergen.html

[6] JaDis - http://pbil.univ-lyon1.fr/software/jadis.html

Other cases, and particularly those where inconsistencies are present are also carefully noted.

Presently, this data gathering procedure applies exclusively to genes, but we intend to extend it to other genomic elements, first of all to anonymous markers that are mapped in many map types (this would allow the user to directly compare different maps types), but also biologically informative sites such as satellites, breakpoints, CpG islands etc.

3   What Kind of System for Comparative Mapping?

A system managing comparative mapping knowledge must allow comparisons of marker position in several maps, but also the mapping of biological properties (G+C content, expression...) and their inter specific comparisons. This implies constraints both on data representation and querying capabilities. Very few databases are dedicated to comparative mapping. Present systems offer different levels of integration of mapping information. We can mention the Animal Genome database in Japan [16], which focuses on comparative gene mapping among pig, cattle, mouse and human. This database contains information concerning genes related to phenotypic traits and any genetic and cytogenetic localization known. Graphical map display is also available. It allows a one locus access, though precluding global analysis. Homology relations between genes are stipulated in the database, however interspecific links between data are not always symmetrical. There is also MICADO [17], a database that manages bacterial genomic information. MICADO links eubacterial and archaebacterial sequences, genetic maps and information on mutants. A graphical interface allows the parallel comparison of genomes. Hyperlinks are created as a complement, to reference other general and specialized related information resources.

Databases based on the ACE model [18] such as AceDB [18] or AtDB [20] allow the comparison of different maps within a species. To a lesser extent, GDB, MGD or Bovmap [21] can also be cited. All of them are in fact specific databases, though they should not be considered as comparative mapping oriented databases. However, in addition to links to other data resources, they contain modules to compare maps among species.

Most of present systems thus use a simplified representation of mapping data; queries are mainly limited to intuitive navigation through information. The more complete system (ACE) is the only one allowing recursive representation of mapping (a map becomes a marker in a larger map) but remains monospecific. Examining the querying capabilities is the simplest way to point out limitations of present systems. None of them are able to answer "simple" requests such as:

" list pairs of homologous genes localized on the murin chromosome 2 and the human chromosome 1"

or

" get all human, complete genes sequences with a known orthologous gene in mouse, which has a sequenced neighboring gene at less than 1 cM"

Moreover, user-controlled inferences have to be applied to generate the complete answer to these questions. For example, reasoning can enrich one type of map from another to produce a more complete answer (see Fig. 1). Reasoning can also apply to inference of markers themselves, for example repetitive elements. User control is also necessary here, particularly when no consensual definition exists, e.g. CpG islands.

Finally, the user should find analysis tools that will enable him to answer biological questions. Let us take an example: the study of regional recombination rates. The idea is to exhibit the features of chromosomal regions that present variations in the frequency of the recombination events. To answer this question, genetical distances are to be expressed as a function of physical position in order to obtain a recombination profile along the chromosome. Presently both physical and genetical localization are not associated to genes, but to anonymous fragments of DNA. To associate genic information to recombination frequencies, mapped genes have to be added to the recombination profile. Finally, the analysis of sequences associated to genes makes it possible to study the variations of recombination levels along the chromosome in relation to genic contextual features.

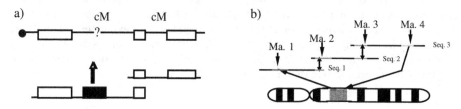

Fig. 1. Non explicit information can be extracted from positional data. (a) - Both physical (bottom) and genetic (top) localizations may be available for some genes or markers, but when there is only a physical localization, it can be used to infer the genetic localization. Because the relation between genetic and physical maps is not linear, such inference is not obvious. In order to make it more clear, a preliminary modeling operation is required to establish the relation between physical and genetic mapping information. (b) - Local connection of nucleotide sequences can also be used to infer element localization, although this implies taking into consideration orthologous elements. Consequently, considering mapping information, as well as evolutive relations, is a prerequisite to inferring positional information from sequence appendings.

4   What Kind of Model for Comparative Mapping?

As seen above, complex information of very different types must be managed. The complexity results from the diversity of the information, but also from the huge number of interrelations between this information. The chosen model must take into account both type complexity and the central role played by relations.

RDBMS (Relational Data Base Management Systems) manage relations efficiently, so they can produce strong and reliable data models. Nevertheless, the relational model is not adapted to complex data schemes. Conceived for tabulated data, it cannot express the complexity of inter-relations of biological objects (see the evolution of the GDB data model [22]).

ACEDB is a model dedicated to genomic mapping in one species. As we have seen, its query capabilities and information representation are not sufficient for comparative mapping. More generally, the developments needed appear to be too large to construct a new dedicated model, even if these models are always more efficient in the context of their development (see ACEDB or ACNUC [23] for sequence management).

OODBMS (Object Oriented Data Base Management Systems) particularly fit for representations of biological concepts: they allow a coupling of the data and the available operators, thus producing a relatively intuitive model of the biological entities represented. Computational object manipulation mimics biological behavior. As genomics is evolving due to huge genome mapping and sequencing projects, data management systems must be flexible and capable of easily integrating new fields of knowledge. Object systems have such capabilities, allowing the construction of modular and hierarchical structures, and thus providing real adaptability to complex data. Nevertheless, relations cannot be clearly defined.

In order to express relations within an object oriented system, we designed a conceptual data scheme based on UML (Unified Modeling Language) definitions. This is an object oriented data model that integrates and expresses relations as peculiar objects. Ongoing research at INRIA Rhône-Alpes on AROM [24], an implementation of an UML-like modeling language as a knowledge representation system, is being used for GeMCore development.

5   GeMCore, a Knowledge Base Dedicated to Mapping Mammalian Genomes

Avoiding redundancy is crucial because redundancy produces inconsistency (when the information to be corrected is recorded many times, some recordings may go uncorrected). Inconsistency management is even more difficult if this information lies in different data structures. Hierarchical structuring of data gives a unique representation of information shared by different elements of the system, thus avoiding the multiplication of storage spaces. Furthermore, when peculiar structures associated to specific information are defined, it is easier to evaluate inconsistency than if the same type of data were stored in heterogeneous structures.

In the GeMCore data model, three major hierarchy classes have been defined (Fig. 2):

- the *Element* hierarchy: it structures a collection of genomic elements such as genes, satellites or transposable elements. This hierarchy separates the elements that have a biological reality from artificial elements produced by genome projects (e.g. clones or expressed sequence tags - EST). Biological elements are defined according to their function. Non-informative elements such as anonymous DNA are separated from functional elements. Genes are distinguished from functional elements that are not transcribed. *Pseudogenes* which are defined at the same level as *Genes,* inherit from *Genes_Related* elements.

- the *Map* hierarchy: three sub-types are defined. Two of them (*Physical_map* and *Genetical_map*) correspond exactly to the biological concepts they represent. The third type named *Set* concerns maps that contain sets of objects that do not have any exact localization. This last group of maps is particularly useful to represent

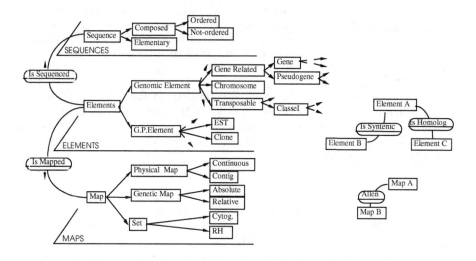

Fig. 2. Simplified scheme of the GeMCore data model. An entity, for example a gene, can be seen as a functional element, a marker on a map or a sequence depending on the objective of the user. In GeMCore, these different levels of information are described in three hierarchy classes: *Sequence*, *Map* and *Element* (left). To express the links existing between these three kinds of objects, two associations (inserts) were defined. *Is_Mapped* and *Is_sequenced* join maps and sequences to the biological element respectively. Associations specific to some data levels are also defined (right). Evolutive relations (*Is_Homolog*) are defined for biological elements. The *Is_syntenic* association expresses that elements are on the same chromosome, whereas *Allen* association allows reasoning on the relative position of maps

    cytogenetic bands or hybrid radiation segments. It also allows users to define their own sets of elements with regard to the question they want to answer (for example, the map of disease genes).

- the *Sequence* hierarchy: it organizes sequences through a hierarchy that distinguishes elementary from non elementary sequences. *Elementary* sequences are made of a unique nucleic acid fragment, while *Composed* sequences correspond to entries that describe collections of fragments that are more or less ordered (currently the unfinished genome sequences from the academic human genome project consortium).

Some complementary classes such as *Species*, *Protein* or *Information*, have been defined in addition to the hierarchies described above.
    Basically, two kinds of relations are defined in the model:

- Comparison relations express evolutive relations (Fig. 3) or similarity relations depending on the objects that are linked (biological objects or nucleotide sequences).

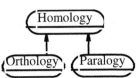

Fig. 3. Homology relations can be specialized. Orthology, which links two genes that result from a speciation event, can be distinguished from paralogy that links elements deriving from a duplication event. Both naturally inherit from the homology relation.

-    Positional relations defined by Allen algebra [25]. Allen introduced a formalism in order to represent intervals by their ends (that is two coordinates on one axis) and suggested the use of constraint satisfaction to reason on these relations. In GeMCore, these relations are structured in a hierarchy (Fig. 4).

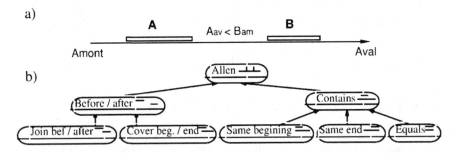

Fig. 4. (a) - According to the Allen formalism, for the A gene to be before the B gene, it is sufficient that the upper coordinate of B be bigger than the lower coordinate of A. (b) - With his formalism, Allen defines thirteen binary relations between interval pairs. In the GeMCore Model, these relations are summarized and structured into a hierarchy where the lower the hierarchy level, the more precise the features of the relative positions of the related elements.

As for classes, certain relations lie outside the hierarchies (*codes_for, is_product_of, belongs_to_species*...).

The three hierarchy classes have been defined to facilitate legibility, making it easier to detect inconsistencies and make updates.

Since data are separated between three independent structures, one hierarchy can be updated independently of other data structures. This is particularly interesting when data come from different sources that are not updated at the same time.

However, the model splits information concerning a given biological entity into three hierarchies, so that it no longer exists as a single entity within the system. The biological notions, the sequence data and the localization relative to a gene belong to a unique entity, which has to be expressed in the model. This is done by "corresponding" relations defined between the biological object, the map and the sequence (Fig. 5).

6   Conclusion

It highly important to associate all the biological data available in order to extract new information. Our goal was at first to define a generic data model allowing not only

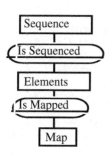

Fig. 5. *GenomicElement*, Sequence and Map classes are tightly linked: as for a sequence, one or more map can be associated to a gene. Depending on the goals of the user, the gene can be seen through its biological properties, its sequence or its localization. Thus it is necessary to express the links existing between all the objects describing the same biological entity. These links are modeled in GeMCore through the relations *Is_Sequenced* and *Is_Mapped*.

data management, but also comparative analysis of mapping of biological and molecular data. GeMCore is an entity-association model designed to perform object comparisons, as well as evaluation and comparison of the relations between the objects in the database. This can be done thanks to the explicit representation of these relations. Our model takes into account the constraints inherent to the domain of comparative mapping and includes many data types (maps, location data, sequence phenotypes, homology...), thus making it valuable for many fields of biology (molecular biology, molecular evolution, medicine, agronomy). A concrete confrontation of the model with real data and analysis will be required for validation. We are thus working on a sub-model in the laboratory of Biométrie et Biologie Evolutive in Lyon using the AROM knowledge representation system. This knowledge base contains human and mouse data. It will be coupled to GeMME and should include other mammalians in the future.

Our problematic deals with the comparison of mapping data, and appears to be more complex than simple construction of data management systems. The goal is to produce a tool that will be able to manage profusely diverse data, while keeping in mind the biological nature of the information contained in the data to be processed. This goal implies the use of consensual terms and methodologies. When building the GeMCore data model we had to precisely define many concepts used in the field of comparative mapping and, more broadly, in methodology dedicated to the management and processing of data produced by genome projects. The GeMCore data scheme could also be helpful in defining an ontology dedicated to comparative genomic mapping.

References

1. Andersson, L., Archibald, A., Ashburner, M., Audun, S., Barendse, W., Bitgood, J., Bottema, C., Broad, T., Brown, S., Burt, D., Charlier, C., Copeland, N., Davis, S., Davisson, M., Edwards, J., Eggen, A., Elgar, G., Eppig, JT., Franklin, I., Grewe, P., Gill, T. 3RD, Graves, J.A., Hawken, R., Hetzel, J., Womack, J., et *al.*: Comparative genome organization of vertebrates. The first international workshop on comparative genome organization. Mamm. Genome **7** (1996) 717-734

2. Rouault, J.P., Kuwabara, P.E., Sinilnikova, O.M., Duret, L., Thierry-Mieg, D., Billaud, M.: Regulation of dauer larva development in *Caenorhabditis elegans* by daf-18, a homologue of the tumour suppressor PTEN. Current Biology **9** (1999) 329-332

3. Spataro, B., Gautier, C.: Analyse comparative des genomes humains et murins avec l'interface GeMME. Recueil des actes, 1ères Journées Ouvertes : Biologie, Mathématiques, Informatique, 3-5 mai 2000, Montpellier - G. Caraux, O. Gascuel, M.F. Sagot (eds.) (2000) 55-64

4. Bernardi, G.: Isochores and the evolutionary genomics of vertebrates. Gene **241** (2000) 3-17

5. Matassi, G., Sharp, P., Gautier, C.: Chromosomal location effects on gene sequence evolution in mammals. Current Biology **9** (1999) 786-791

6. Letovsky, S.: GDB: integrating genomic maps. In: S. Letovsky (ed.) Bioinformatics: Databases and Systems. Kluwer Academic Publishers, Boston, Dortrecht, London (1999) 85-98

7. Eppig, J.T., Richardson, J.E., Blake, J.A., Davisson, M.T., Kadin, J.A., Rinwald, M.: The mouse genome database and the gene expression database: genotype to phenotype. In: S. Letovsky (ed.) Bioinformatics: Databases and Systems. Kluwer Academic Publishers, Boston, Dortrecht, London (1999) 119-128

8. Collins, A., Frezal, J., Teague, J., Morton, N.E.: A metric map of humans:23,500 loci in 850 bands. Proc. Natl. Acad. Sci. USA **93** (1996) 14771-14775

9. Scott, A.F., Amberger, J., Brylawski, B., McKusick, V.A.: OMIM: online mendelian. inheritance in man. In: S. Letovsky (ed.) Bioinformatics: Databases and Systems. Kluwer Academic Publishers, Boston, Dortrecht, London (1999) 77-84

10. Achard, F., Cussat-Blanc, C., Viara, E., Barillot, E.: The new Virgil database: a service of rich links. Bioinformatics **14** (1998) 342-348

11. Pruitt, K.D., Katz, K.S., Sicotte, H., Maglott, D.R.: Introducing RefSeq and LocusLink: curated human genome resources at the NCBI. Trends Genet. **16** (2000) 44-47

12. Thompson, J.D., Higgins, D.G., T.J., Gibson.: CLUSTALW , improving the sensitivity of progressive multiple alignment through sequence weighting, position-specific gap penalties and weight matrix choice. Nucl. Acid. Res **22** (1994) 4673-4680

13. Altschul, S.F., Gish, W., Miller, W., Myers, E.W., Lipman, D.J.: Basic local alignment search tool. J Mol Biol. **215** (1990) 403-410

14. Altschul, S.F., Madden, T.L., Schaffer, A.A., Zhang, J., Zhang, Z., Miller, W., Lipman, D.J.: Gapped BLAST and PSI-BLAST: a new generation of protein database search programs. Nucleic Acids Res. **25** (1997) 3389-3402

15. Duret, L., Perrière, G., Gouy, M.: HOVERGEN: comparative analysis of homologous vertebrate genes. In: S. Letovsky (ed.) Bioinformatics: Databases and Systems. Kluwer Academic Publishers, Boston, Dortrecht, London (1999) 21-36

16. Wada, Y., Yasue, H.: Development of an animal genome database and its search system - Comput. Applied Biosci. **12** (1996) 231-235

17. Biaudet, V., Samson, F., Bessieres, P.: Micado – a network oriented database for microbial genomes – Comput. Applied Biosci., **13** (1997) 431-438

18. Thierry-Mieg, J., Thierry-Mieg, D., and Stein, L.: ACEDB: The ACE Database Manager. In: S. Letovsky (ed.) Bioinformatics: Databases and Systems. Kluwer Academic Publishers, Boston, Dortrecht, London (1999) 265-278

19. Durbin, R., Thierry-Mieg, J.: The ACeDB genome database. In: Suhai, S. (ed.) Computational Methods in Genome Research, Plenum Press,New-York (1994) 45-55

20. Flanders, D.J., Weng, S., Petel, F.X., Cherry, J.M.: AtDB, the Arbidopsis thaliana database, and graphical web display of progress by the Arbidopsis genome initiative. Nucleic Acids Res. **26** (1998) 80-84

21. http:// locus.jouy.inra.fr/cgi-bin/bovmap/intro2.pl

22. Fasman, K.H., Letovsky, S.I., Cottingham, R.W., Kingsbury, D.T.: Improvements to the GDB human genome database - Nucleic Acids Res. **24** (1996) 57-63

23. Gouy, M., Gautier, C., attimonelli, M., Lanave, C., di Paola, G.: ACNUC – a portable retrieval system for nucleic acid sequence databases: logical and physical designs and usage. CABIOS **1** (1985) 167-172

24. Page, M., Gensel, J., Capponi, C., Bruley, C., Genoud, P., Ziebelin, D.: Representation des connaissances au moyen de classes et d'associations: le système AROM. Conf. langage et modèles à objet (LMO'00), Mt Saint-Hilaire, CA (2000) 91-106
25. Allen, J.F.: Maintaining knowledge about temporal intervals – Comm. of the ACM **26** (1983) 832–843

Appendix: Markers, Maps, and Recombination

The genome is made of several chromosomes. Each of them includes a DNA molecule, which is a linear polymer of four nucleotides (adenine, guanine, cytosine, thymine). Genetic information is coded in these nucleotide sequences.

Proteins are molecules that ensure most of the cell functions. The information necessary to build proteins is stored in protein genes. A gene is a subsequence of the genome. Expression of its information content requires two successive processes: transcription that generates a mRNA molecule (the messenger) and translation that converts this RNA molecule into protein. Unexpectedly, only a small fraction of the genome contains genes. In mammals, about 95% of genome has no genetic information content.

A marker is any genomic object on the genome that can be identified non ambiguously: a gene is a marker as well as an anonymous sequence of a genomic fragment. Genome maps appear mainly as an ordered series of markers. A set of markers located on the same chromosome is said to be syntenic.

Several distances between markers can be determined. When the complete genome is known, the number of nucleotides between two markers is known. This is the physical distance between the two markers. When the sequence is not known, the physical distance can be estimated from experimental data. Another important distance is the genetic distance. Each individual has two sets of homologous chromosomes, one from his father, the other from his mother. During gamete production, pairs of homologous chromosomes cross over, exchanging sequences. The genetic distance can be determined from the expected number of cross-overs between two markers (the recombination rate). Comparing physical and genetical distance is useful to address important biological points in various fields of biology including population genetics, medical research, and evolution.

Optimal Agreement Supertrees

David Bryant

LIRMM, 161 rue Ada, 34392 Montpellier, France Cedex 5
bryant@lirmm.fr

Abstract. An *agreement supertree* of a collection of unrooted phyloge-
netic trees $\{T_1, T_2, \dots, T_k\}$ with leaf sets $\mathcal{L}(T_1)$, $\mathcal{L}(T_2), \dots, \mathcal{L}(T_k)$ is an
unrooted tree T with leaf set $\mathcal{L}(T_1) \cup \cdots \cup \mathcal{L}(T_k)$ such that each tree
T_i is an induced subtree of T. In some cases, there may be no possible
agreement supertrees of a set of trees, in other cases there may be exponentially many. We present polynomial time algorithms for computing an
optimal agreement supertree, if one exists, of a bounded number of binary
trees. The criteria of optimality can be one of four standard phylogenetic criteria: binary character compatibility; maximum summed quartet
weight; ordinary least squares; and minimum evolution. The techniques
can be used to search an exponentially large number of trees in polynomial time.

1 Introduction

The general phylogenetic supertree problem is how to combine phylogenies for
different sets of species into one large "super"-phylogeny that classifies all of the
species. The problem is motivated by physical, computational, and statistical
constraints on the construction of large scale phylogenies.

In this paper we present a supertree method for combining a bounded number
of binary, unrooted, phylogenetic trees. We search for supertrees such that every
input tree is a restriction of the supertree to a subset of the set of species. When
multiple supertrees exist we choose the supertree that is optimal with respect to
one of four standard phylogenetic optimization criteria. Even though the number
of possible supertrees can be exponentially large, we can determine an optimal
supertree in polynomial time.

1.1 Agreement Supertrees

We begin with a few definitions. An **unrooted phylogenetic tree** T is a finite,
acyclic connected graph with no internal vertices of degree two and degree one
vertices (leaves) labelled uniquely by members of a finite leaf set $\mathcal{L}(T)$. A tree T'
is an **induced subtree** of T if T' can be obtained from T by deleting vertices
and edges and supressing vertices of degree two. Intuitively T' represents the
restriction of T to a subset of the leaf set. If T' is an induced subtree of T with
leaf set $L' \subseteq \mathcal{L}(T)$ then we write $T' = T|_{L'}$. An **agreement supertree** of a
collection of trees $\{T_1, T_2, \dots, T_k\}$ with leaf sets $\mathcal{L}(T_1), \mathcal{L}(T_2), \dots, \mathcal{L}(T_k)$ is an

O. Gascuel, M.-F. Sagot (Eds.): JOBIM 2000, LNCS 2066, pp. 24–31, 2001.

unrooted tree T with leaf set $\mathcal{L}(T_1) \cup \cdots \cup \mathcal{L}(T_k)$ such that each tree T_i is an induced subtree of T.

The agreement supertree is the dual of the **agreement subtree**, recalling that S is an agreement subtree of trees $T_1, \ldots, T_k$ if S is an induced subtree of each of $T_1, \ldots, T_k$ (c.f [3,6]). A collection of trees $T_1, \ldots, T_k$ on leaf set L are agreement supertrees of a collection of trees $S_1, \ldots, S_l$ on subsets of L if and only if $S_1, \ldots, S_l$ are agreement subtrees of $T_1, \ldots, T_k$, and every leaf in L appears in at least one S_i.

There can be exponentially many agreement supertrees for a collection of trees $\mathcal{T}$, even when the number of trees in $\mathcal{T}$ is bounded [4]. When there is more than one agreement supertree we select one that is optimal according to one of four standard phylogenetic optimization criteria. The choice of the four optimization criteria is determined by algorithmic constraints - we discuss the possible extension to other criteria in Sect. 6. Note that the optimization criteria require *global* information for the entire set of taxa present in the input trees. Hence the algorithms are better tailored to divide and conquer applications, rather than the assembly of different data sets.

Though the number of agreement supertrees can be exponentially large, the algorithms determine an optimal agreement supertree in polynomial time, provided that the number of input trees is bounded. The agreement supertree methods presented here are practical for a small number of large input trees, rather than a large number of small input trees.

Finally we note that agreement supertree methods do not handle conflict in the set of input trees. If two input trees resolve the relationship between species in a different way then there will be no agreement supertrees for the collection. There are numerous possible *ad hoc* solutions to this shortcoming (e.g. removing leaves involved in conflicts, modifying input trees). However the more general problem of how to systematically resolve conflict in supertrees is still hotly debated within phylogenetics. At this early stage we present our results as tools to be incorporated into practical supertree methods once some of the fundamental systematic questions have been answered.

Despite the false modesty, the algorithmic results presented in this paper have several direct applications, a few of which we discuss in Sect. 5. In particular we stress the ability to search over an exponentially large number of trees in polynomial time.

2   Four Standard Optimization Criteria

In this section we describe the phylogenetic criteria used for optimization. Discussion of the motivations, use, and characteristics of the various criteria can be found, for example, in [10]. Here we give only sufficient detail to describe the problems.

2.1   Binary Character Weights

Given a phylogenetic tree T, removing an edge of T induces a bipartition of the leaf set of T, called a **split** of T. The set of all splits of T is denoted $splits(T)$, and a single split is written as $A|B$, using the standard notation for partitions.

A split can be viewed as a **binary character** where each leaf is assigned a value of 0 or 1 depending on which block of the split it is in. Given a weighted collection of binary characters we can score a tree T by summing the weight of the characters corresponding to splits in T. The total summed weight is called the **binary character compatibility score** of T. We wish to find T that maximizes the binary character compatibility score.

We will assume that all binary characters have non-negative weights.

2.2   Quartet Criteria

A (resolved) **quartet** is an unrooted binary tree on four leaves. There are three possible quartets for each set of four leaves. For any tree T the **quartet set** of T, denoted $q(T)$, is the set of all quartets that are induced subtrees of T.

Suppose that we have assigned a weight $w(ab|cd)$ for every possible quartet $ab|cd$ such that $\{a, b, c, d\} \subseteq \mathcal{L}(T)$. The **quartet score** of a tree T is the sum of the weights $w(ab|cd)$ over all $ab|cd \in q(T)$. We want to find a tree with maximum quartet score.

We will assume that all quartets have non-negative weights.

2.3   Least Squares (OLS)

Suppose that T is an unrooted tree and the edges of T are given real valued weights. The **leaf to leaf distance** between any two leaves of T is the sum of the edge weights along the path between the two leaves. In this way we can construct a distance matrix p containing all of the leaf to leaf distances between leaves in T. The **sum of squares** distance between p and a given distance matrix d for $\mathcal{L}(T)$ is defined

$$\|p - d\|^2 = \sum_{x \in \mathcal{L}(T)} \sum_{y \in \mathcal{L}(T)} (d_{xy} - p_{xy})^2 \ . \tag{1}$$

For a fixed tree T the edge weights for T that give a distance matrix p minimizing $\|p - d\|^2$ are called the **OLS edge length estimates** for T.

The **ordinary least squares (OLS) score** of a tree T is the value $\|p - d\|^2$ given by the OLS edge length estimates. We want to find a tree T with minimum OLS score.

2.4   Minimum Evolution (ME)

The minimum evolution criteria is closely related to OLS. Given an unrooted tree T and a distance matrix d we first calculate the OLS edge estimates for T

from d. The **minimum evolution (ME) score** for T is then the sum of these edge weights. The minimum evolution score is usually only evaluated for binary trees[1]. We want to find a tree with minimum ME score.

3 Split Constrained Phylogenetic Optimization

In this section we describe the results of [1,2,5,7] that provide the machinery for the optimal agreement supertree algorithms. The results all follow the same theme: the introduction of split based constraints that make phylogenetic reconstruction polynomial time solvable. More formally we have:

INSTANCE: Set $\mathcal{S}$ of splits on leaf set L. Degree bound d.
PROBLEM: Is there a tree T with degree bound d such that $splits(T) \subseteq \mathcal{S}$? If so, find the tree(s) with degree bound d and $splits(T) \subseteq \mathcal{S}$ with

1. maximum binary compatibility score (with respect to a given weighting of splits in $\mathcal{S}$). [1]
2. maximum quartet score (for a given quartet weighting). [7]
3. minimum OLS score (for a given distance matrix). [5]
4. minimum ME score (for a given distance matrix). [2,5]

All of these versions of the problem can be solved in polynomial time (in the number of splits and leaves), when d is bounded by a constant.

4 Optimal Agreement Supertree Algorithms

4.1 Split Set Constraints

We show how the split constrained optimization algorithms can be used to construct optimal agreement supertrees.

Let $\mathcal{T} = \{T_1, \ldots, T_k\}$ be a collection of trees. Put $L_i = \mathcal{L}(T_i)$ for each i and $L = L_1 \cup \cdots \cup L_k$. Define

$$\mathcal{S}(\mathcal{T}) = \left\{ A_1 \cup \cdots \cup A_k | B_1 \cup \cdots B_k : \begin{array}{l} A_i | B_i \in splits(T_i) \cup \{\emptyset | L_i\}, i = 1, 2, \ldots, k \\ (A_1 \cup \cdots \cup A_k) \cap (B_1 \cup \cdots B_k) = \emptyset \end{array} \right\} f. \tag{2}$$

Note that we will assume $A_i | B_i \in splits(T_i)$ implies $B_i | A_i \in splits(T_i)$. Put $n = |L|$. There are at most $2n - 3$ splits in each tree so $|\mathcal{S}(\mathcal{T})|$ is $O(2^k n^k)$. Furthermore

Theorem 1. *Let T be an unrooted phylogenetic tree with leaf set L. If each tree $T_i \in \mathcal{T}$ is an induced subtree of T then $splits(T) \subseteq \mathcal{S}(\mathcal{T})$.*

---

[1] It would be interesting to determine whether or not optimal minimum evolution trees can be assumed to be binary

Proof. Suppose that each tree $T_i \in \mathcal{T}$ is an induced subtree of T. Thus $T_i = T|_{L_i}$ and

$$splits(T_i) = splits(T|_{L_i}) \tag{3}$$
$$= \{A \cap L_i | B \cap L_i : A|B \in splits(T)\} - \{L_i | \emptyset\}. \tag{4}$$

Hence for each $A|B \in splits(T)$ we have

$$A \cap L_i | B \cap L_i \in splits(T_i) \cup \{L_i | \emptyset\} \ . \tag{5}$$

It follows that

$$A|B = (A \cap L_1) \cup \cdots \cup (A \cap L_k)|(B \cap L_1) \cup \cdots \cup (B \cap L_k) \tag{6}$$

and $A|B$ belongs to $\mathcal{S}(\mathcal{T})$. □

We now take advantage of the fact that the input trees are all binary.

Lemma 1. *If $\mathcal{T}$ contains only binary trees and there is an agreement supertree T of $\mathcal{T}$ then there is a binary agreement supertree T' of $\mathcal{T}$ such that $splits(T) \subseteq splits(T')$.*

Proof. Suppose that T is an agreement supertree of $\mathcal{T}$ that is not binary. Clearly there exists a binary tree T' such that $splits(T) \subseteq splits(T')$. We will show that T' is also an agreement supertree for $\mathcal{T}$.

For each $T_i \in \mathcal{T}$ we have $T_i = T|_{L_i}$ and, since $splits(T) \subseteq splits(T')$, we also have $splits(T|_{L_i}) \subseteq splits(T'|_{L_i})$. The tree T_i is binary, so $splits(T_i) = splits(T|_{L_i}) \subseteq splits(T'|_{L_i})$ implies $T_i = T'|_{L_i}$, completing the proof. □

We now have all the machinery needed for the main result.

Theorem 2. *Let $\mathcal{T} = \{T_1, \ldots, T_k\}$ be a collection of binary trees with leaf sets $L_1, \ldots, L_k$. We can determine whether an agreement supertree of $\mathcal{T}$ exists in $O(4^k n^{2k+1})$ time. If there is an agreement supertree for $\mathcal{T}$ then we can find the agreement supertree with optimum*

1. *binary character compatibility score (with respect to a given weighting of splits) in $O(4^k n^{2k+1})$ time.*
2. *summed quartet weight (for a given quartet weighting) in $O(2^k n^{k+4} + 4^k n^{2k+1})$ time.*
3. *least squares score (for a given distance matrix) in $O(8^k n^{3k})$ time.*
4. *ME score (for a given distance matrix) in $O(8^k n^{3k})$ time.*

Proof. By Theorem 1 we have that any agreement supertree T for $\mathcal{T}$ satisfies $splits(T) \subseteq \mathcal{S}(\mathcal{T})$. With all of the optimization criteria (except perhaps ME, which is usually defined only for binary trees) a tree T' with $splits(T) \subseteq splits(T')$ will have at least as good score as T. Hence, using Lemma 1, we can assume that the optimal agreement supertrees are binary.

The problem therefore reduces to finding an optimal binary tree contained within a given collection of splits. Applying the methods stated in Sect. 3 is is possible to obtain the given complexities. □

We note that, when the number of trees k is bounded by a constant, we have *polynomial time* algorithms for determining optimal agreement supertrees. When k is unbounded, the problem is NP-complete (by a reduction from QUARTET COMPATIBILITY [9]).

5   Applications to Optimization

To illustrate how these methods may be used during optimization we describe two applications.

5.1   Divide and Conquer

Given two (or more) binary trees we now have a method for combining them in a way that optimizes phylogenetic criteria. The most obvious application of these algorithms is as the basis of a divide and conquer algorithm.

There are several possibilities for ways to subdivide the data set. The DCM method [8] uses distance data to divide the sequences into closely related clusters. However the merge algorithms we describe here will work well even if all of the input trees contain leaves that are widely scattered throughout the combined tree. For this reason, the subdivision process appears less critical. We propose the use of a random subdivision, biased to produce subsets that have close to equal sizes.

5.2   Testing Stability

The standard technique for phylogenetic optimization is to conduct a tree search: an initial tree is constructed; Neighbouring trees are examined, where 'neighbouring' means within one branch swap or NNI exchange of the original tree; If a better tree is found, this is chosen as the next tree and the search continues. At each step at most $O(n)$ or $O(n^2)$ trees are searched.

We can use the merge algorithms presented in this paper to search an exponentially large number of trees without a dramatic increase in complexity. We randomly divide the set of leaves into two parts, calculate the two induced subtrees, and then calculate an optimal agreement supertree for these two trees. An agreement supertree will always exist since the leaf sets of the subtrees are disjoint. If we have found a global optimum then this procedure will always produce the original tree. If it finds a better tree we take this new tree and repeat the process.

6   Discussion

In this abstract we have outlined a method for constructing an optimal agreement supertree of a set of binary trees, if such an supertree exists. The algorithm can search an exponentially large collection of trees (the possible agreement supertrees) in polynomial time. Our results lead to several directions of further research:

6.1 Extension to Other Criteria

It would be useful to determine whether the algorithms can be extended to optimize over other phylogenetic criteria such as parsimony score or likelihood. In both cases we are pessimistic. With parsimony, the structure of one part of a tree can alter how best to resolve the structure of another part of the tree, making parsimony less suitable for dynamic programming algorithms such as those we use here. With maximum likelihood it is unknown whether the likelihood of a *single* tree can be computed in polynomial time! In both cases, the above results can be used to make a rough search of the tree space that would precede a more detailed, and computationally expensive, local search.

6.2 Extension to Non-binary Trees

The next direction of future research would be to determine if the algorithms can be extended to handle non-binary trees. The problem is that we require a degree bound in order to use the algorithms outlined in Sect. 3. We suspect that an appropriate modification of the algorithms will give a polynomial time algorithm (for a bounded number of trees) even without a degree bound. We have derived the required modifications for the special case when the leaf sets of the input trees are disjoint, but the the complexity of the general problem remains open.

6.3 Trees with No Agreement Supertree

At the moment the algorithms will simply terminate if there is no agreement supertree possible. This occurs when, for example, there are conflicts between the input trees. It would be interesting to extend the algorithm to derive a consensus supertree method capable of handling input trees with conflicts. First, however, we must address the question of how to handle conflict between different trees. The subtree merger algorithm of [8] simply contracts edges causing problems. This will, in general, lead to poorly resolved trees. Once suitable criteria for combining subtrees are defined, it should be possible to extend the optimal agreement supertree algorithms above to give a general optimal subtree merger technique.

Acknowledgements. I thank the anonymous referees for their helpful comments and suggestions. The paper was written while I was a postdoctoral fellow based in the laboratory of Olivier Gascuel.

References

1. Bryant, D.: Hunting for trees in binary character sets. Journal of Computational Biology **3(2)**, (1996) 275-288.

2. Bryant, D.: Hunting for trees, building trees and comparing trees: theory and method in phylogenetic analysis. Ph.D. thesis, Dept. Mathematics, University of Canterbury (1997).

3. Finden, C.R. and Gordon. A.D. Obtaining common pruned trees. Journal of Classification, **2** (1986) 255–276.

4. Gordon, A.D. Consensus supertrees: the synthesis of rooted trees containing overlapping sets of leaves. Journal of Classification, **3** (1986) 335–348.

5. Bryant, D. Fast algorithms for constructing optimal trees with respect to OLS and minimum evolution criteria. In preparation.

6. Goddard, W., Kubicka, E., Kubicki, G., and McMorris, F.R.. (1994). The agreement metric for labelled binary trees. Mathematical Biosciences **123** 215–226.

7. Bryant, D. and Steel, M. Constructing optimal trees from quartets. Journal of Algorithms (to appear)

8. D.H. Huson, S. Nettles, and T. Warnow: Obtaining highly accurate topology estimates of evolutionary trees from very short sequences. Proc. 3rd Annual Int. Conf. Comp. Mol. Biol. (RECOMB), (1998) 198–209.

9. M. Steel: The complexity of reconstructing trees from qualitative characters and subtrees. Journal of Classification, **9** (1992) 91-116.

10. Swofford, D. L., G. J. Olsen, P. J. Waddell and D. M. Hillis: Phylogenetic Inference. in D. M. Hillis, C. Moritz, and B. K. Mable, eds. Molecular Systematics, second edition. Sinauer, Sunderland, Mass (1996) 407–514.

Segmentation by Maximal Predictive Partitioning According to Composition Biases

Laurent Guéguen

CEB-LIS – UPMC Paris VI
gueguen@biomserv.univ-lyon1.fr

Abstract. We present a method for segmenting qualitative sequences, according to a type of composition criteria whose definition and evaluation are founded on the notion of predictors and additive prediction. Given a set of predictors, a partition of a sequence can be precisely evaluated. We present a language for the declaration of predictors. One of the problems is to optimize the partition of a sequence into a given number of segments. The other problem is to obtain a suitable number of segments for the partitioning of the sequence. We present an algorithm which, given a sequence and a set of predictors, can successively compute the optimal partitions of the sequence for growing numbers of segments. The time- and space-complexity of the algorithm are linear for the length of sequence and number of predictors. Experimentally, the computed partitions are highly stable regard to the number of segments, and we present an application of this approach to the determination of the origins of replication of bacterial chromosomes.

Broadly speaking, the aim of the classification process is to optimize the dividing-up a set of objects into classes, in line with given criteria, so as to identify a structure, if any such exist.

Once a set is organized into classes, the next aim may be to obtain a description of these classes, which would provide a high-level redescription of the set. This description could be set either after or during the classification process. But it may be very difficult to obtain "good" descriptions of existing classes, so the approach we adopted was to simultaneously describe and construct classes, and to evaluate the quality of the classification according to the accuracy of the descriptions.

In this paper, we outline a method for partitioning sequences into described segments in such a way that the descriptions could be used to evaluate the quality of the corresponding segments. We first present the overall approach, and then an efficient, optimal partitioning algorithm. Finally, we link this method to the problem of partitioning genomic sequences according to relevant composition biases.

O. Gascuel, M.-F. Sagot (Eds.): JOBIM 2000, LNCS 2066, pp. 32–44, 2001.
© Springer-Verlag Berlin Heidelberg 2001

1 Prediction

1.1 Prediction of Sequences

We use Gower's concept of **prediction** [3]: the quality of a class is measured by the ability of an "optimal" predictor belonging to a given set of predictors to predict the properties of the elements of the class in question. For sequences of letters, a property of a given position is the letter or word that is readable at this position. For example, to say that the predictor of a sequence of letters is the letter A means that each letter of this sequence should be an A, and the resulting high-level description of this sequence is: "We can read an A at each position in the sequence". This description will be more or less accurate, according to the proportion of A in the sequence. To say that the predictor of a sequence is the word 'AT' means that the word 'AT' should be readable at each position in the sequence, but it is clear that such a prediction can be correct for, at most, half the positions in the sequence.

The notion of prediction can be extended to entire sequences. For example, to say that the predictor of a class is A with a factor 0.3, and T with a factor 0.7, means that we predict: "We can read an A at 30 % and a T at 70 % of the positions of the sequence". We call this predictor [A(0.3)T(0.7)] (see Sect. 1.2).

Let δ be a predictor, $\mathcal{S}$ a sequence, and s_i a position in this sequence. The **prediction** of δ at s_i is termed $\pi_\delta(s_i) \in \mathbb{R}$.

For $\delta = $ A, if the letter at position s_i is A, $\pi_\delta(s_i)$ is 1, otherwise it is 0. For $\delta = $ 'AT', if the word at position $s_i s_{i+1}$ is AT, $\pi_\delta(s_i)$ is 1, otherwise 0. For $\delta = $ [A(0.3)T(0.7)], if the letter at s_i is A, $\pi_\delta(s_i)$ is 0.3; if the letter is T, $\pi_\delta(s_i)$ is 0.7; otherwise, it is 0.

The prediction of δ on a segment $\mathcal{S}$ is the sum of the predictions of δ for all the positions on this segment. We call this prediction $\pi_\delta(\mathcal{S})$:

$$\pi_\delta(\mathcal{S}) = \sum_{s_i \in \mathcal{S}} \pi_\delta(s_i)$$

In a geometrical representation, predictions for a sequence may be considered as a scalar product. In Fig. 1, for instance, we represent the sequence as a walk on the orthonormal basis (e_A, e_T), starting at the origin. If we read an A (resp. T) at the first position, the first step in the walk is e_A (resp. e_T). In a given representation, the vector corresponding to sequence $\mathcal{S}$ is called the **sequence vector of** $\mathcal{S}$, noted as $v_\mathcal{S}$. In the same way, a predictor δ is also represented by a **predictor vector**, v_δ. In this context, the prediction of the predictor δ on the sequence $\mathcal{S}$ is the scalar product of v_δ and $v_\mathcal{S}$, termed $v_\delta.v_\mathcal{S}$.

For a set of predictors $\mathcal{D}$, the optimal predictors of $\mathcal{S}$ are those that provide the best prediction on $\mathcal{S}$, i.e. those whose vectors provide the best scalar product with the sequence vector of $\mathcal{S}$. The corresponding optimal prediction is $\pi_\mathcal{D}(\mathcal{S})$:

$$\pi_\mathcal{D}(\mathcal{S}) = \max_{\delta \in \mathcal{D}} \pi_\delta(\mathcal{S})$$

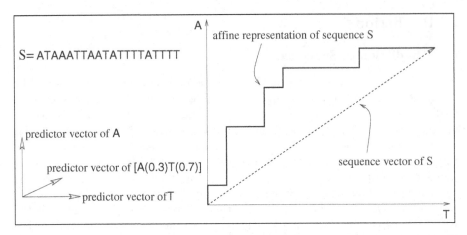

Fig. 1. Vectorial representation of a sequence, and of predictors A, T and [A(0.3)T(0.7)].

There is competition between the different predictors of $\mathcal{D}$. Following the geometrical representation, and in order to ensure that the predictor chosen among the set $\mathcal{D}$ is the one whose vector is the most colinear with the sequence vector, we have to normalize the different predictor vectors. For example, if $\mathcal{D}$ is the set {A, T, [A(0.3)T(0.7)]}, the normalization is carried out by multiplying the predictions of [A(0.3)T(0.7)] by 1.313 (see Fig. 2). This predictor is termed [A(0.3)T(0.7)](1.313).

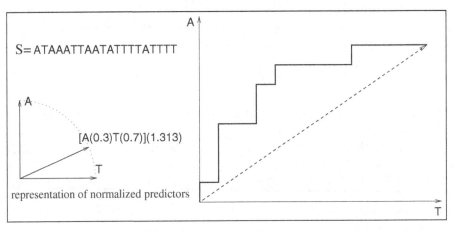

Fig. 2. Vectorial representation of normalized predictors. The predictor [A(0.3)T(0.7)](1.313) is the one that best fits the sequence S.

1.2   Language of Predictors

We constructed a language that would allow us to declare different predictors. We use the word "**lexicon**" to designate a set of words belonging to this language which determines the set $\mathcal{D}$ of the available predictors. In a lexicon, the words are separated by **spaces**. Each word corresponds to a predictor, according to the rules given below. Now, the two terms "predictor" and "word" are equivalent.

- If α is a letter in $\mathcal{S}$, α is in the language, and its prediction function is the characteristic function, i.e.[1]:

$$\pi_\alpha(s_i) = \mathbb{1}_{\alpha=s_i} = \begin{vmatrix} 1 \text{ if the letter in position } s_i \text{ is } \alpha, \\ \text{otherwise } 0; \end{vmatrix}$$

- The sign '?' is in the language, its prediction function is:

$$\pi_?(s_i) = 1$$

- If p is a predictor and $x \in \mathbb{R}$, p(x) is a predictor belonging to the language and:

$$\pi_{\mathtt{p}(x)} = x.\pi_\mathtt{p}$$

- If $\mathtt{p}_1, \ldots, \mathtt{p}_k$ are predictors, $[\mathtt{p}_1\mathtt{p}_2\ldots\mathtt{p}_k]$ is a predictor belonging to the language, and:

$$\pi_{[\mathtt{p}_1\mathtt{p}_2\ldots\mathtt{p}_k]} = \sum_{i=1}^{k} \pi_{\mathtt{p}_i}$$

- If $\mathtt{p}_1, \ldots, \mathtt{p}_k$ are predictors, '$\mathtt{p}_1\mathtt{p}_2\ldots\mathtt{p}_k$' is a predictor belonging to the language, and:

$$\pi_{\text{'}\mathtt{p}_1\mathtt{p}_2\ldots\mathtt{p}_k\text{'}}(s_i) = \pi_{\mathtt{p}_1}(s_i).\mathbb{1}_{(\pi_{\mathtt{p}_2}(s_{i+1})>0)\wedge\ldots\wedge(\pi_{\mathtt{p}_k}(s_{i+k-1})>0)}$$

Combinations within the framework of this syntax make possible the declaration of a very large range of predictors. For instance, the word '$\mathtt{A?C}$'(2) stands for a predictor whose prediction is 2 at positions where the letter is A and the letter which occurs two positions after it is C, 0 at all other positions. The predictor $[\text{'}\mathtt{A[CA]G}\text{'}(-2)\text{'}\mathtt{CA}\text{'}](3)$ gives a prediction of -6 at positions where the word is ACG or AAG, 3 where it is CA, otherwise 0.

2   Predictive Partitioning of Sequences

2.1   Predictive Partitions

A **partition** of a sequence is a division of the sequence into separate segments such that their union equals the whole sequence. We call **caesura** the junction

---

[1] $\mathbb{1}_{\text{true}} = 1$ and $\mathbb{1}_{\text{false}} = 0$

of two consecutive segments. We call a k-**partitions** a partition into k segments, and the set of these k-partitions is termed $\mathbb{P}_k(\mathcal{S})$.

The prediction of a partition follows the same logic as the prediction of a sequence: our aim is to optimize the prediction of the composition of segments, and the prediction of a partition is the sum of the optimal predictions of its segments. Then, if the k-partition P is made up of segments $\mathcal{S}_1, \mathcal{S}_2, \ldots, \mathcal{S}_k$, its prediction by $\mathcal{D}$ is:

$$\pi_{\mathcal{D}}(P) = \sum_{j=1}^{k} \pi_{\mathcal{D}}(S_j) = \sum_{j=1}^{k} \max_{\delta \in \mathcal{D}} \pi_{\delta}(S_j)$$

For example, given the sequence

$$\text{ATAAATTAATATTTTATTTT}$$

and the lexicon A T [A(0.3)T(0.7)](1.313), then for the 3-partition

$$\text{ATAAATTA ATATT TTATTTT}$$

these segments have successively A, [A(0.3)T(0.7)](1.313) and T as optimal predictors, and the prediction of this 3-partition is $5 + 3.55 + 6 = 14.5451$.

The larger the prediction of a k-partition, the better the predictions that can be made about its segments. Given a set of predictors and a number of classes k, our goal is to obtain the optimal division into k segments, i.e. to find the optimal partitions among the entire set of k-partitions. The evaluation of the maximal k-partitions of $\mathcal{S}$ according to the predictors set $\mathcal{D}$ is termed $M_k^{\mathcal{D}}(\mathcal{S})$.

$$M_k^{\mathcal{D}}(\mathcal{S}) = \max_{(\mathcal{S}_1, \ldots, \mathcal{S}_k) \in \mathbb{P}_k(S)} \sum_{j=1}^{k} \max_{\delta \in \mathcal{D}} \pi_{\delta}(S_j)$$

In the example given above, an optimal 3-partition is

$$\text{ATAAA TTAATATTTTA TTTT}$$

These segments have successively A, [A(0.3)T(0.7)](1.313) and T as optimal predictors, and the corresponding prediction is 16.0093.

In terms of geometrical representation, an optimal k-partition is one in which the vectors representing the segments are most colinear to their best predictors among $\mathcal{D}$ (see Fig. 3). Finding a "good" partition means finding a compromise between very precise descriptions of only a few of the k segments and a good average of all the k segments. Because the prediction is additive in the neighborhood of a junction between two segments, the caesura is set in such a way that the prediction of the maximum elements is relevant. Given a set of predictors, and considering this geometrical representation, we can say that a sequence has a structure when its affine representation can be divided up into segments whose vectors are distinctly colinear with one of the predictors.

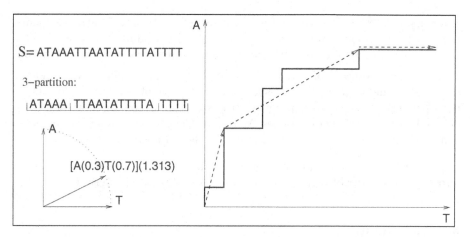

Fig. 3. Vectorial representation of a 3-partition. Dashed arrows represent the vectors of the segments of this partition. We can see that the vector of predictor A is the most colinear with the first arrow, [A(0.3)T(0.7)](1.313) with the second, and T with the third.

2.2   Number of Segments

A priori, it is reasonnable to ask an experimenter to define a problem in terms of a sequence and a set of predictors[2]. But the precise number of segments which will successfully partition a sequence cannot be known in advance. So, from the point of view of classification, and with the hypothesis of a structure in a sequence, it may be useful to estimate the number of segments that "best" discloses this structure. And if there is no structure at all, it should be possible to demonstrate this. In fact, more than one class may be relevant, as in the case of nested clusters. There are numerous ways to estimate the number(s) of classes in a set, and we will not list them all, but a number of them can be found in [2]. However, none of them is perfect.

A "good" summary of the composition of the sequence implies the "right" number of classes. Yet by definition, the optimal prediction does not decrease with the number of segments, and the global maximum is reached when all the elements have been separated, which is not a very interesting solution. The "right" number of classes must be found: if there are too few, the high-level re-description obtained will be too vague; if there are too many, this redescription will be too complicated. One way to assess the relative usefulness of the diffe-rent numbers of classes is to compute the optimal k-partitions for successive k between 1 and a given value. We call this process **partitioning**, and we give the same name to the result of such a process.

A partitioning can be used to study the optimal prediction in terms of the number of classes. Qualitatively, a sharp bend in the curve of the prediction is the sign of a relevant number of segments because, for a lower number of

---

[2] We shall discuss the choice of this set in the Sect. 4.

segments, the prediction will be too vague, and for a higher number, the gains in prediction will be outweighed by the increase in the numbers of classes. Such a point corresponds to a good balance between summarization and description.

A further advantage of partitioning a sequence is that it may reveal another structure of the sequence, i.e. the successive revelation of segments according to ·their prediction by a "good" predictor, the structure becoming more and more detailed with the number of classes. Moreover, the composition of the segments is not subject to a priori limitations, such as that of a threshold.

We next present an algorithm for computing the partitioning of a sequence into a given number of segments given a sequence and a set of criteria [4].

3   Algorithm

There already exists an algorithm for computing the optimal partitionings of sequences, for any kind of evaluation function of the classes [5]. This algorithm is based on Fisher's lemma [1]): if $\{S_1, S_2, \ldots, S_k\}$ is an optimal k-partition of $\cup_{i=1}^{k} S_i$, then $\{S_1, S_2, \ldots, S_{k-1}\}$ is an optimal $k - 1$-partition of $\cup_{i=1}^{k-1} S_i$. The time-complexity of this algorithm is quadratic with the size of the sequence, and this is critical for very long sequences, such as genomic sequences.

In the context of prediction, the algorithm we are introducing here uses the previous lemma and another observation: given a set of predictors $\mathcal{D}$, all the positions of a class are predicted by the same predictor of $\mathcal{D}$ (whether this prediction is correct or not). The time- and space-complexity of the algorithm are linear with the length of the sequence, which makes possible the partitioning of very long sequences.

Let $s_1, s_2, \ldots, s_n$ be the elements of sequence $\mathcal{S}$, whose size is n.
Given $i \geqslant j$, we call $\langle i, j \rangle$ the set $\{i, i+1, \ldots, j-1, j\}$ and
$$s_{\langle i,j \rangle} \quad \text{the subsequence} \quad s_i \ldots s_j .$$

For each $\delta \in \mathcal{D}$, and for $i \geqslant k$, if we compute, for all the k-partitions of the subsequence $s_{\langle 1,i \rangle}$, the prediction such that the segment containing s_i is predicted – forcibly – by the predictor δ, we call $M_k^\delta(s_{\langle 1,i \rangle})$ the maximum of these predictions.

The aim is to compute $M_k^{\mathcal{D}}(s_{\langle 1,n \rangle})$ for each k, and to backtrack the corresponding optimal partitions. In an optimal k-partition, the segment containing s_i is predicted by an optimal predictor:

$$M_k^{\mathcal{D}}(s_{\langle 1,i \rangle}) = \max_{\delta \in \mathcal{D}} M_k^\delta(s_{\langle 1,i \rangle})$$

Fisher's lemma is used to compute $M_k^\delta(s_{\langle 1,i \rangle})$. An optimal k-partition of $s_{\langle 1,i \rangle}$ for which the optimal predictor of the segment containing s_i is δ conforms to at least one of the two possibilities:

— either s_{i-1} and s_i belong to the same segment, in which case the segment containing s_{i-1} is predicted by δ,

$$\overset{s_1\ s_2\ \ldots}{\underset{\textstyle}{\left|\rule{0pt}{1em}\right.}} \quad\dotfill\quad + \quad + \quad \dotfill\quad \overset{\ldots s_{i-1}\ s_i}{\underset{\ldots\delta\quad\ \delta}{\left|\rule{0pt}{1em}\right.}}$$

and the prediction is: $M_k^\delta(s_{\langle 1,i\rangle}) = \pi_\delta(s_i) + M_k^\delta(s_{\langle 1,i-1\rangle})$

— or s_{i-1} and s_i belong to two different segments, in which case the kth segment is $\{s_i\}$, and the other $k-1$ segments together constitute an optimal $(k-1)$-partition of $s_{\langle 1,i-1\rangle}$,

$$\overset{s_1\ s_2\ \ldots}{\underset{\textstyle}{\left|\rule{0pt}{1em}\right.}} \quad\dotfill\quad + \quad + \quad \dotfill\quad \overset{\ldots s_{i-1}\ \ s_i}{\underset{\ldots\delta'\quad\ \delta}{\left|\rule{0pt}{1em}\right.}}$$

and the prediction is: $M_k^\delta(s_{\langle 1,i\rangle}) = \pi_\delta(s_i) + M_{k-1}^{\mathcal{D}}(s_{\langle 1,i-1\rangle}).$

The maximum of these two formulae is:

$$M_k^\delta(s_{\langle 1,i\rangle}) = \pi_\delta(s_i) + \max\left(M_{k-1}^{\mathcal{D}}(s_{\langle 1,i-1\rangle}), M_k^\delta(s_{\langle 1,i-1\rangle})\right),$$

and, as there is no k-partition of $s_{\langle 1,k-1\rangle}$,

$$M_k^\delta(s_{\langle 1,k\rangle}) = \pi_\delta(s_k) + M_{k-1}^{\mathcal{D}}(s_{\langle 1,k-1\rangle})$$

To sum up:

$$\boxed{\begin{array}{l}
\forall i, M_1^{\mathcal{D}}(s_{\langle 1,i\rangle}) = \max_{\delta\in\mathcal{D}} \pi_\delta(s_{\langle 1,i\rangle}) \\[1.2em]
\forall\delta, \forall k, M_k^\delta(s_{\langle 1,k\rangle}) = \pi_\delta(s_k) + M_{k-1}^{\mathcal{D}}(s_{\langle 1,k-1\rangle}) \\[1.2em]
\forall\delta, \forall k, \forall i > k, M_k^\delta(s_{\langle 1,i\rangle}) = \pi_\delta(s_i) + \max\left(M_{k-1}^{\mathcal{D}}(s_{\langle 1,i-1\rangle}), M_k^\delta(s_{\langle 1,i-1\rangle})\right) \\[1.2em]
\forall k, \forall i \geqslant k, M_k^{\mathcal{D}}(s_{\langle 1,i\rangle}) = \max_{\delta\in\mathcal{D}} M_k^\delta(s_{\langle 1,i\rangle})
\end{array}}$$

So if, for a given k, the set $\{M_{k-1}^{\mathcal{D}}(s_{\langle 1,i\rangle}), i \geqslant k-1\}$ is preserved, we can compute successively, for $i \geqslant k$, the set of the $M_k^{\mathcal{D}}(s_{\langle 1,i\rangle})$ using all the $M_k^\delta(s_{\langle 1,i\rangle})$. We then compute successively the partitioning $M_1^{\mathcal{D}}(\mathcal{S})$, $M_2^{\mathcal{D}}(\mathcal{S})$, ..., $M_k^{\mathcal{D}}(\mathcal{S})$. Backtracking this computation gives an optimal k-partition for each k.

The computation of these optimal partitions is linear over time: for each $e \in \langle 2, k\rangle$, for each $i \in \langle e, n\rangle$, for each $\delta \in D$, $M_e^\delta(s_{\langle 1,i\rangle})$ is computed in a constant time using $M_e^\delta(s_{\langle 1,i-1\rangle})$ and $M_{e-1}^{\mathcal{D}}(s_{\langle 1,i-1\rangle})$. Moreover, for each $i \in \langle e, n\rangle$, the computation of $M_e^{\mathcal{D}}(s_{\langle 1,i\rangle})$ requires a comparison between $|\mathcal{D}|$ elements. Then, at each iteration of the number of segments, the computing time approximates to $\sum_{i=e}^{n} |\mathcal{D}| = |\mathcal{D}|.(n - e + 1)$. In conclusion, $M_k^{\mathcal{D}}(\mathcal{S})$ is computed with a time-complexity of $O(k.|\mathcal{S}|.|\mathcal{D}|)$. This algorithm can thus be used on very long sequences.

For a given k, there may in fact be several maximally predictive k-partitions, and it may be useful to calculate all these, especially if they have very different caesuras. With adjacency lists, it is possible to find the set of all the equally optimal k-partitions without any increase in complexity.

4   Applications

4.1   Composition Biases in Sequences

Genetic sequences are subject to numerous functional and structural constraints, and there are composition biases between their different parts. For us, "composition bias" means that there is a significant difference between different parts of a sequence for a measurable criterion, e.g. the comparison between the proportion of G and the proportion of C on a DNA string. In a number of bacteria the expression of this C--G bias is related to the replication process, due to the fact that these proportions are very different between the leading and the lagging part of a DNA double-strand sequence [6]. In [7], a linear discriminant analysis method is used to find, and especially, to evaluate the relative merits of different criteria for distinguishing between the leading and lagging parts of a bacterial chromosome; these criteria can be based on the C--G bias or the frequency of codons in genes. The aim of detecting such biases is to facilitate the study of biological data, though the explanation of those biases raises new problems for biological research.

We plan to link the notion of composition bias to that of prediction. To arrive at an objective criterion of composition bias, we use the notion of prediction, and that of language as defined in Sect. 1.2. Our second aim is to compute such partitionings, using the algorithm presented in Sect. 3.

If a composition bias can be translated into a prediction, using this approach, we can partition the sequence into segments that optimally mirror this bias.

Taking a geometrical point of view, we consider a composition bias $\mathcal{B}$ such that, in an affine representation of a sequence, which has been set up according to the parameters of $\mathcal{B}$, each trend τ_i of the bias is related to a transect T_{τ_i} of the space. This means that if the vector of a segment is in T_{τ_i}, the segment follows τ_i. At the boundary between two transects, the segment follows the two trends to the same extent. Looking at all the trends in a bias, we have a partitioning of the space into transects $\{T_{\tau_1}, \ldots, T_{\tau_p}\}$ (as in Fig. 4). Looking at a set of predictors $\{\delta_1, \ldots, \delta_n\}$, for each i let T_{δ_i} be the set of vectors v such that $\forall j \neq i, v.v_{\delta_j} < v.v_{\delta_i}$. T_{δ_i} is a transect of the space. In order to express bias $\mathcal{B}$ using the prediction, we have to find p predictors δ_i such that $\{T_{\delta_1}, \ldots, T_{\delta_p}\} = \{T_{\tau_1}, \ldots, T_{\tau_p}\}$. This problem can be solved easily by using the fact that at the boundary between two transects, the prediction of a segment is the same for the two corresponding predictors.

The algorithm and the declaration of the language used to compute optimal predictive partitionings have been programmed. The computation of partitionings turns out to be very efficient: for example, on a Gateway computer with a

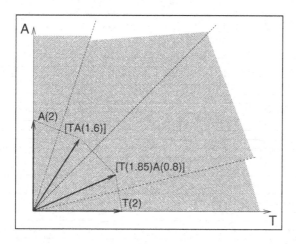

Fig. 4. Predominance transects of the predictors of the lexicon A(2) [TA(1.6)] [T(1.85)A(0.8)] T(2). The striped areas are the transects, which are related to the trends in the studied bias.

Pentium III processor and 256 Mb of RAM, taking a sequence of 2.10^6 letters and four predictors, the construction of an optimal partition into 100 segments takes less than ten minutes. In this way, it has been possible to study the partitioning process extensively.

4.2   Experiments

The discovered segments are very stable with the numbers of segments. For instance, the partitioning of the sequence shown in the Fig. 5 almost gives the impression of a hierarchical structure. This is not in fact the case, but it reinforces the impression that partitioning reveals hidden structures, since it seems to bring out increasingly detailed features of the composition of the sequence. Visually, such figures do indeed give a thorough view of the intricacy of the expression of the different tendencies of a bias within a given sequence. Moreover, the stability means that the determination of a "suitable" number of segments becomes less critical, because the caesuras remain, along with several numbers of classes.

We partitioned several bacterial genomic sequences according to the C--G imbalances in different parts of these sequences. We wanted to check whether the bias noted by Lobry [6] made it possible to determine the origin and terminus of replication. To express this bias, we used the C G lexicon . The chromosomes of the bacteria studied are circular, so we dicided them up into linear sequen-ces. Figure 6 gives the result of partitioning *Haemophilus Influenzae* into ten segments. We see that the leading and lagging parts are the first to be extrac-ted. Figure 7 shows a very sharp bend in the prediction curve in terms of the number of classes, at 3 segments, revealing that the most significant expression

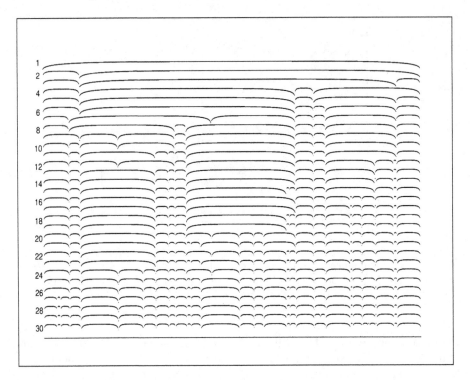

Fig. 5. Example of the partitioning of a sequence. The sequence is the line at the bottom. Each class within a partition is represented by an arc. The number of classes increases from top to bottom.

of the G--C bias is exactly correlated with the leading and lagging parts of the chromosome.

Table 1 shows, for a number of bacteria and archae-bacteria, the limits of the segments of the 2- and 3-partitions, using the same C G lexicon, for which there is a sharp bend in the prediction curve. The selected number of classes is either 2 (which means that the origin of replication is at position 0, i.e. at the position where the circular chromosome has been split) or 3.

It seems clearly that this method could be extensively used for determining the composition of sequences such as genomes and proteins. The large range of criteria according to which partitionings can be carried out may be useful tool in biological research, e.g. in determining of the relation between the replication process and other criteria (cf [7]), the study of isochores, or that of variations in codon usage inside a chromosome. And, as this method is precise and objective, it is a good way to evaluate the relevance of particular criteria, relating partitions to knowledge about the properties of particular sequences.

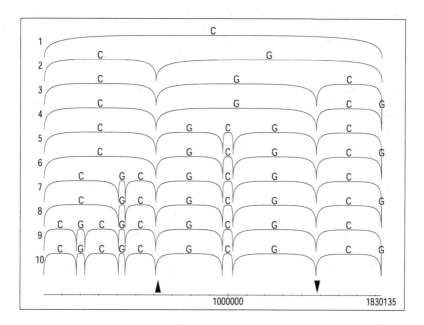

Fig. 6. Partitioning of *H. influenzae* into 10 classes, using a C G lexicon. The chromosome, which is circular, is represented as a linear sequence (the graduated line). The black triangles represent the locations of the origin (pointing upwards) and terminus (pointing downwards) of replication.

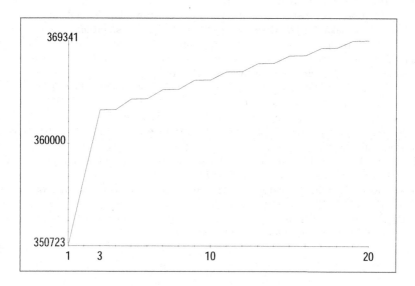

Fig. 7. Prediction in terms of the number of classes of partitioning into twenty classes of *H. influenzae*, using the C G lexicon.

Table 1. Origins and termini of replication and with the caesurae found by partitioning, in sequences for which an increase in prediction shows a sharp bend. The positions are given as percentages of the overall length of the sequence. "Caesura of C towards G" means that the predictor C (resp. G) is on the left (resp. right) of the caesura.

Organism	Origin	Terminus	Caesura of C towards G	Caesura of G towards C
B. burgdorferi	50.1	0	50.3	0
B. subtilis	0	47.8	0	46.1
C. trachomatis	69	19	69.1	18.7
E. coli	84.6	34.1	84.6	33.4
H. influenzae	32.9	80.1	33.2	80.1
H. pylori	96	47	98	48.8
M. tuberculosis	0	49	0	46.3
M. pneumoniae	25.1	73	25	74.1
M. genitalium	0	?	0	50.8
M. thermoautotrophicum	75	24	74.9	23.3
P. horikoshii	?	?	0	43.6
R. prowazekii	0	56	0	56.5
T. pallidum	0	48	0	48.6

References

1. W.D. Fisher. On grouping for maximal homogeneity. *Journal of the American Statistical Association*, 53:789–798, 1958.
2. A.D. Gordon. Cluster validation. In C. Hayashi, N. Ohsumi, K. Yajima, Y. Tanaka, H.H. Bock, and Y. Baba, editors, *Studies in Classification, Data Analysis, and Knowledge Organization: Data Science, Classification, and Related Methods*, pages 22–39, Kobe, March 1996. IFCS, Springer-Verlag. http://www-solar.dcs.st-and.ac.uk/~allan/.
3. J.C. Gower. Maximal predictive classification. *Biometrics*, 30:643–654, 1974.
4. L. Guéguen, R. Vignes, and J. Lebbe. Maximal predictive clustering with order constraint: a linear and optimal algorithm. In A. Rizzi, M. Vichi, and H. Bock, editors, *Advances in Data Science and Classification*, pages 137–144. IFCS, Springer Verlag, July 1998.
5. D.M. Hawkins and D.F. Merriam. Optimal zonation of digitized sequential data. *Mathematical Geology*, 5(4):389–395, 1973.
6. J.R. Lobry. Asymmetric substitution patterns in the two dna strands of bacteria. *Mol. Biol. Evol.*, 13(5):660–665, 1996.
7. E.P.C. Rocha, A. Danchin, and A. Viari. Universal replication biases in bacteria. *Molecular Microbiology*, 32(1):11–16, 1999.

Can We Have Confidence in a Tree Representation?

Alain Guénoche[1] and Henri Garreta[2]

[1] IML - CNRS
[2] LIM - Université de la Méditeranée,
163 Av. de Luminy, 13288 Marseille Cedex 9

Abstract. A tree representation distance method, applied to any dissimilarity array, always gives a valued tree, even if the tree model is not appropriate. In the first part, we propose some criteria to evaluate the quality of the computed tree. Some of them are metric; their values depend on the edge's lengths. The other ones only depend on the tree topology. In the second part, we calculate the average and the critical values of these criteria, according to parameters. Three models of distance are tested using simulations. On the one hand, the tree model, and on the other hand, euclidean distances, and boolean distances. In each case, we select at random distances fitting these models and add some noise. We show that the criteria values permit one to differentiate the tree model from the others. Finally, we analyze a distance between proteins and its tree representation that is valid according to the criteria values.

1 Introduction

In this paper we consider aspects of the tree representation of a given proximity measure on a finite set X. This measure can be a distance, satisfying the metricity conditions, or a simple dissimilarity evaluated from any description of the X elements. Briefly speaking, we only have an array D, with dimension $n = |X|$, that is symmetrical with null values on the diagonal. The tree reconstruction methods define an X-tree having X as set of leaves and its internal nodes connect exactly three edges with non negative length. [Barthélemy, Guénoche 1991].

When X is a set of homologous biological sequences and when the distance values are proportional to the number of mutations, the tree model is justified by the evolution theory of Darwin. But when this distance measures something else, such as a structural similarity or a functional proximity, it is not certified that an underlying tree exists and that the D values permit to infer this tree.

There are many other domains where X-trees are calculated from a proximity measure, such as Cognitive Psychology, for studies about perception - noises, smells, etc. In Phylogeny, the tree nodes represent *taxa* or common ancestors, but in Psychology, these nodes correspond to *categories* and the reality of a tree organization of these categories is a hypothesis that must be verified and justified.

O. Gascuel, M.-F. Sagot (Eds.): JOBIM 2000, LNCS 2066, pp. 45–56, 2001.

The aim of this paper is to study the appropriateness of the tree model to represent a given distance D. To do that, we apply a tree reconstruction method to get a tree A and its associated distance D_a. Then, we measure the gap between D and D_a using metric and topological criteria. To evaluate how large this gap is, we realize simulations to estimate the observed values of the criteria when the initial distance is close to a tree distance and to determine confidence intervals. If the criteria values for D belong to these intervals, A and the tree model are accepted. But if not, one can try another reconstruction method, but if none of these gives an acceptable tree, the model will be finally abandoned. Then one can check if other classical models, such as euclidean or boolean models are suited.

Similar studies have been realized previously. We refer to Prutzansky, Tversky and Caroll [1982] who try to determine if a given distance is better fitted with a hierarchical clustering method or with a multidimensional scaling one, that is if D is closer to the ultrametric model than to the euclidean one. This study extends their work, dealing with the tree model that is more general, adding the boolean model, and offering the possible conclusion that none of these models is appropriate.

In Sect. 2, we recall some criteria to evaluate the quality of an X-tree compared to a distance on X. There are two kinds of criteria, metric and typological. The first ones come from Statistics, evaluating differences between the D and D_a values. The second ones are less classical. They only depend on the tree topology and are independent from the edges length. These criteria also permit one to compare different trees obtained by several reconstruction methods. Even if they do not agree, they give quantitative arguments to prefer one representation.

In Sect. 3, we describe our procedures to generate random distances considering three models: on the one hand the tree distances, and on the other hand, the euclidean distances and the boolean distances established from binary attributes. For the tree distances we fix the ratio between the lengths of the internal and external edges. For the other distances, several dimensions of the spaces are tested. Realizing simulations, we evaluate average and critical values. The comparisons of the corresponding tables show that the tree model cannot be confused with the other distance models.

In Sect. 4, we study a TAT proteins family (Twin Arginine Pathway) for which we build a tree that is finally validated by the method described here, according to the tables.

2   Criteria to Evaluate X-Trees

Let D be a $n \times n$ dissimilarity array such that for all $\{x, y\} \in X^2$, $D(x, y) = D(y, x)$ and for all $x \in X, D(x, x) = 0$. Let A be an X-tree given by any reconstruction method. The tree A, together with an edge-weighting allows one to calculate a tree distance such that for all $\{x, y\} \in X^2$, $D_a(x, y)$ is equal to the length of the path in A between x and y.

To evaluate the quality of A, we compare the two distances D and D_a. This is realized using a program *Qualitree* working with two files: one for the tree, registered according to the Newick format and the other one for the distance D having the Phylip format. This program can be obtained at http://biol10.biol.umontreal.ca/guenoche/.

2.1   Metric Criteria

Among all the criteria proposed in Statistics, we have selected:
    - the *distortion*, which is the average of the percentage of differences:

$$Dis = \frac{2}{n(n-1)} \sum_{x<y \in X} \frac{|D(x,y) - D_a(x,y)|}{D(x,y)}$$

    - the *stress*, which corresponds to the square root of the quadratic difference divided by the average distance value:

$$St = \sqrt{\frac{\sum_{x<y \in X}(D(x,y) - D_a(x,y))^2}{\sum_{x<y \in X} D(x,y)^2}}$$

    - the *variance accounted for*, which corresponds to the quadratic difference divided by the variance of the distance (Dm is the average of the D values):

$$Vaf = 1 - \frac{\sum_{x<y \in X}(D(x,y) - D_a(x,y))^2}{\sum_{x<y \in X}(D(x,y) - Dm)^2}$$

Other criteria such as the correlation coefficient or the square root of the quadratic difference, are also calculated by the program. As they are not used here, in our analysis, they are not discussed further.

2.2   Topological Criteria

Generally D is not a tree distance, but the four point condition [Buneman 1971] gives information about the tree topology of any quadruple in the underlying tree. If, in this hypothetical tree, elements x and y are separated from elements z and t by at least one edge, then:

$$D(x,y) + D(z,t) < Min\{D(x,z) + D(y,t), D(x,t) + D(y,z)\} \tag{1}$$

If D is a tree distance the two largest sums are equals. When $D(x,y)+D(z,t)$ is the smallest of the three sums, D indicates topology T_1 as figured in Fig. 1. If it is another sum that is the smallest, then D indicates either T_2 or T_3. If the three sums are equal it corresponds to the unresolved topology T_0.

Since D indicates a topology for any quadruple, one can compute two criteria:

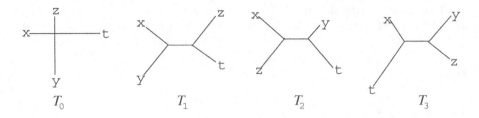

Fig. 1. The four possible tree topologies for quadruple $\{x, y, z, t\}$

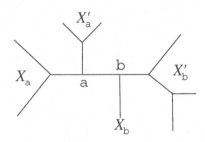

Fig. 2. The four parts of X corresponding to edge (a, b)

- the rate of *well designed quadruples*, these are quadruples having the same topology according to D and D_a.

$$R_q = \frac{1}{\binom{n}{4}}|\{x, y, z, t\} \text{ with the same topology for } D \text{ and } D_a|.$$

- the rate of *elementary* well designed quadruples, which we define as follows.

Let (a, b) be an internal edge in the tree. It defines four subtrees, corresponding to the directions on both sides of this edge. According to Fig. 2, let X_a, X'_a, X_b and X'_b be the sets of leaves belonging to these four subtrees. If this edge is correct, any quadruple such that $x \in X_a, y \in X'_a, z \in X_b, t \in X'_b$ must verify condition (1). In that case, we shall say that $\{x, y, z, t\}$ supports the edge (a, b). Our second topological criteria is the percentage of elementary quadruples that supports a internal edge. It only counts quadruples that are difficult to recover since the two pairs are separated by a unique edge.

$$R_e = \frac{\sum_{(a,b)\in A} |\{x,y,z,t\} \text{ supporting } (a,b)|}{\sum_{(a,b)\in A} |X_a||X'_a||X_b||X'_b|}$$

Let $R_e(a, b)$ be the rate of elementary quadruple supporting the edge (a, b). It lies between 0 and 1 and it indicates a quality measure for this edge, the reliability of which increases with the rate. This criterion has, for the distance methods, a role similar to the bootstrap one, commonly used in phylogeny when starting from sequences, but that cannot be used from a simple distance array. This criterion can also validate edges having low bootstrap value, because sequences share very few sites corresponding to this edge.

Our last topological criterion does not depend on a particular tree, nor the employed method. It indicates the tendency for a distance to be representable by a tree. In their article in 1982 about the comparisons of tree versus euclidean representations, Prutzansky, Tversky and Caroll concluded that an arboricity criterion was a good indicator of the best model. It was the percentage of triangles having the length of the medium side closer to the largest side than to the smallest one. In this way, ultrametric distances get an optimal value, but it is not true for all the tree distances, since any metric triangle has an exact tree representation. So we have modified their criterion, considering quadruples.

- An *arboricity* coefficient, which we define as follows.

Let $Smin$, $Smed$ and $Smax$ be the three sums involved in the four point condition and ranked in the increasing order. Since for tree distances the two greatest sums are equal, we measure the percentage of quadruples for which the middle sum is closer to the largest one than to the smallest one.

$$Arb = \frac{1}{\binom{n}{4}}|\{x,y,z,t\} \text{ such that } Smax - Smed < Smed - Smin|$$

3   Evaluating Average and Critical Values

The criteria values are calculated starting from random distances. To select at random distances close to one model, we begin with a valued tree or a set of points in an euclidean or a boolean space and we calculate the initial distance $Dini$. Then we add some noise according to a parameter τ. For each value $Dini(x, y)$, we select uniformly at random a value ε such that $0 \leq \varepsilon \leq \tau$, and we apply formula $D(x, y) := (1 \pm \varepsilon)Dini(x, y)$ in which $\pm$ is also the result of an equiprobable selection: half the values are increased and half the values are decreased with a percentage bounded by τ. The result is a dissimilarity, since the triangle inequality is not necessarily preserved. Then, applying a tree reconstruction method, we obtain a valued X-tree and the associated tree distance D_a. Comparing D and D_a, we evaluate the criteria values on a dissimilarity having a gap to the model corresponding to τ.

For the three models, we use the Neighbor Joining method of Saitou and Nei [1987], with the improvement of Studier and Keppler [1988] to obtain a $O(n^3)$ algorithm. The results given in the following tables correspond to $n = 15$, the number of elements for the protein example. They are computed after 200 trials, using a program MoDist, also available on the same URL. For each criteria, the 200 values realize a distribution from which we calculate the lower (resp. higher) threshold corresponding to 90% of the trials. In other words, the probability to get a value smaller (resp. greater) than the lower (resp. higher) critical value is less or equal to 10%. In the program, a threshold of 5% is also available.

3.1   The Tree Model

To select at random dissimilarities close to a tree distance, we start from a random X-tree. Simulating an ascending process leads to numerous topologies that are rarely obtained [Guénoche and Préa 1998], and some methods give better results with some of them [Gascuel and Jean Marie 1999]. Finally we retain the Yule-Harding procedure [1971] that consists in subdividing any class with more than two elements. This procedure generates a probability distribution on labelled X-trees that are coded in a binary table, each row being an element of X and each column an internal edge of the tree.

Then we give some random weights to the edges in the interval [5, 25]. Because the recovering of the tree is much easier when the internal edges are long, the ratio between the lengths of the internal and the external edges is an important parameter for the simulation process. Let ρ be this parameter ($\rho \leq 1$) ; we multiply by ρ the lengths of the internal edges. It is well known that the symmetrical difference distance between rows of this weighted binary table is a tree distance.

The rate of noise τ applied to distances varies from 5% to 20%. With larger values, the percentages of well designed quadruples can be so weak that the tree is not reliable. The ratio ρ takes values .25, .5 and 1 which give problems that are progressively easier. In Table 1, the average values are printed between the lower and higher critical values. For instance for $\rho = .50$ and $\tau = .15$ the average value for R_q is .93 ; 10% of these distances gives values lower than or equal to .89 and 10% give values greater than or equal to .95.

One can remark that the metric criteria do not much depend on ρ ; in contrast, topological criteria give best results when this ratio increases.

3.2   The Euclidean Model

To appreciate the values given by dissimilarities close to a tree distance, we test distances coming from euclidean and boolean models. For both we must fix the dimension of the space, that is the number p of variables or attributes used to place the X elements. If we want to keep the same number of degrees of freedom as in a tree representation, that is $2n - 3$ (the number of edges), this leads to specific values for p:

- $p = 2$ for euclidean distances (since there are $2n$ random coordinates),
- $p = 1 + \log_2(n)$ for boolean distances (2^p possible different points)

But, a priori, for a given distance we do not know this p value, and we have realized simulations with different possibilities.

To select at random an euclidean distance, we build a table T with n random vectors such that $T(x, k) \in [0, 1]^p$. Then we apply the classical formula:

$$D_e(x, y) = \sqrt{\sum_{k=1}^{p} (T(x, k) - T(y, k))^2}$$

Table 1. Average and critical values for dissimilarities close to tree distances

$$\rho = .25$$

τ	5%			10%			15%			20%		
Dis	1.9	2.1	2.3	3.8	4.2	4.5	5.8	6.3	6.8	7.5	8.4	9.4
St	.02	.03	.03	.05	.05	.06	.07	.08	.09	.10	.10	.11
Vaf	.98	.99	.99	.93	.95	.96	.86	.90	.93	.78	.83	.88
R_q	.96	.98	1	.88	.92	.95	.81	.85	.89	.74	.80	.85
R_e	.91	.96	.99	.75	.83	.88	.64	.71	.77	.57	.63	.70
Arb	.89	.93	.96	.75	.79	.84	.64	.70	.75	.58	.63	.68

$$\rho = .50$$

τ	5%			10%			15%			20%		
Dis	1.9	2.1	2.3	3.8	4.2	4.5	5.7	6.3	6.9	7.7	8.4	9.4
St	.02	.03	.03	.05	.05	.06	.07	.08	.09	.10	.11	.12
Vaf	.99	.99	.99	.95	.96	.97	.90	.92	.94	.83	.87	.90
R_q	.99	1	1	.94	.96	.98	.89	.93	.95	.82	.87	.91
R_e	.98	.99	1	.85	.92	.96	.77	.83	.90	.67	.74	.81
Arb	.95	.97	.99	.84	.88	.91	.75	.79	.84	.67	.71	.76

$$\rho = 1$$

τ	5%			10%			15%			20%		
Dis	1.9	2.1	2.3	3.8	4.2	4.6	5.8	6.4	7.0	7.7	8.4	9.2
St	.02	.03	.03	.05	.05	.06	.07	.08	.09	.10	.11	.12
Vaf	.99	.99	.99	.97	.97	.98	.93	.94	.96	.88	.90	.93
R_q	1	1	1	.97	.98	.99	.93	.96	.98	.89	.92	.95
R_e	.99	1	1	.93	.96	.99	.84	.90	.94	.77	.82	.88
Arb	.97	.99	1	.89	.92	.95	.82	.85	.88	.75	.79	.82

We have retain for p the values $p = 2$, $p \sim \sqrt{n}$, $p = \frac{n}{2}$. As in our simulations $n = 15$, p takes values 2, 4 and 8. In the program, p can take any value lower than or equal to n.

For the euclidean and boolean distances we only apply three values for τ; 0% (the model is perfectly fitted), 10% and 20%. We always indicate the average value (on 200 trials) but only one critical value, the one which leads to a bound between this type of distance and the tree model. For the distortion and the stress criteria that are the lower values, and for the variance accounted for and the topological criteria the higher values. So for $p = 4$ and $\tau = 10\%$, the stress coefficient is lower than or equal to .14 for 10% of the distances, and R_e is greater than or equal .63 for the same percentage.

As it was foreseeable the quality of the metric criteria increases with the number of variables except for Vaf (the variance accounted for). But the topological criteria decreases very quickly. For $p = 2$, R_q has similar values as for tree distances (with $\tau = 20\%$ and $\rho = .25$) and R_e is larger for euclidian distances, but in that case the distortion and the stress coefficients are much larger than the

Table 2. Average and critical values for dissimilarities close to euclidean distances

	τ	0%	10%	20%
$p = 2$	Dis	14.5 (11.1)	15.1 (12.3)	17.1 (14.2)
	St	.18 (.13)	.19 (.15)	.21 (.16)
	Vaf	.80 (.91)	.78 (.90)	.76 (.87)
	R_q	.81 (.88)	.81 (.87)	.80 (.86)
	R_e	.66 (.74)	.65 (.72)	.63 (.71)
	Arb	.76 (.81)	.75 (.79)	.72 (.76)

	τ	0%	10%	20%
$p = 4$	Dis	12.7 (10.8)	13.1 (11.3)	15.1 (13.0)
	St	.15 (.13)	.16 (.14)	.19 (.16)
	Vaf	.72 (.83)	.71 (.80)	.65 (.75)
	R_q	.75 (.80)	.74 (.79)	.72 (.78)
	R_e	.59 (.65)	.58 (.63)	.56 (.61)
	Arb	.67 (.71)	.65 (.69)	.62 (.66)

	τ	0%	10%	20%
$p = 8$	Dis	9.4 (8.4)	10.2 (8.9)	12.1 (10.7)
	St	.12 (.11)	.13 (.11)	.15 (.14)
	Vaf	.67 (.76)	.64 (.73)	.58 (.68)
	R_q	.70 (.75)	.69 (.74)	.66 (.71)
	R_e	.55 (.60)	.54 (.58)	.52 (.56)
	Arb	.61 (.66)	.59 (.63)	.55 (.59)

corresponding values for the tree model. At the opposite, for $p = 8$, the metric criteria, except Vaf that varies as the topological criteria, get similar values as for tree distances having a strong noise, but topological criteria are much lower than for the tree model. So the two types of criteria never agree, and it is always feasible to separate noisy tree distances from euclidean distances, whatever the dimension is.

3.3   The Boolean Model

For boolean distances, each element of X is represented by a binary vector with p components equiprobably equal to 0 or 1. Then we give random weight W to the attributes in the interval $[1,5]$ and we calculate

$$D_b(x, y) = \sum_{k=1}^{p} W(k)|T(x, k) - T(y, k)|$$

We have selected for p the values $p = 1 + \log_2(n)$, $p = \frac{n}{2}$ and $p = n$, that gives integer values 5, 8, 15.

As for euclidean distances, the metric criteria except Vaf are improved when the number of attributes increases, but they never reach values similar to those of tree distances. The topological criteria take similar values as for euclidean

Table 3. Average and critical values for dissimilarities close to boolean distances

		τ	0%	10%	20%
$p = 5$	Dis		20.8 (15.8)	21.8 (17.7)	24.1 (19.8)
	St		.25 (.21)	.26 (.22)	.27 (.23)
	Vaf		.62 (.77)	.62 (.75)	.59 (.73)
	R_q		.76 (.83)	.74 (.80)	.73 (.80)
	R_e		.68 (.78)	.58 (.65)	.57 (.63)
	Arb		.64 (.72)	.67 (.73)	.65 (.70)
$p = 8$	Dis		19.6 (16.9)	20.2 (17.4)	21.6 (18.8)
	St		.23 (.20)	.24 (.21)	.25 (.22)
	Vaf		.54 (.67)	.53 (.65)	.51 (.63)
	R_q		.71 (.77)	.70 (.76)	.70 (.75)
	R_e		.57 (.63)	.54 (.59)	.54 (.58)
	arb		.60 (.66)	.61 (.65)	.59 (.64)
$p = 15$	Dis		18.3 (15.3)	19.1 (16.1)	20.5 (17.1)
	St		.22 (.18)	.23 (.20)	.24 (.21)
	Vaf		.69 (.80)	.67 (.76)	.64 (.74)
	R_q		.71 (.75)	.70 (.74)	.70 (.75)
	r_e		.53 (.58)	.53 (.58)	.52 (.57)
	arb		.61 (.66)	.60 (.64)	.58 (.62)

distances. They can be considered as tree distance values only for $p = 5$ and again topological criteria will make the difference. Finally boolean distances can be distinguished from tree distances.

For these two models, the criteria do not much depend on the rate of noise. Euclidean and boolean distances are very different from a tree distance and to add some noise to the former does not change anything. These distances give a poor tree representation and the quadruples that support some edges seems to be randomly selected.

Finally, to recognize a tree distance, it suffices to observe the metric criteria: the distortion is very low (generally lower than .10) and the stress coefficient lower than .12. The Vaf is greater than .90. These values cannot be obtained by others models except by euclidian distances when the dimension is large. But in that case the topological criteria will make the difference, since in case of possible confusion, tree distances provide values much greater than the observable ones in the euclidian or boolean case.

4   The Protein Example

The twin-arginine translocation (Tat) pathway is capable of export proteins in a folded conformation. This novel protein export system has been revealed in *Escherichia coli*. Five Tat genes have been isolated, and three of them are considered as homologous. For an extended description of the biological questions we refer to Wu et al. [2000].Collaborating with this team, we have studied the evolutionary tree of these genes and a PAM distance have been computed on 15 sequences. There are a few positions that can be aligned on this set and consequently, all the main bootstrap values are not satisfactory, especially for the edge that separate the TatB proteins from the TatA and TatE cluster.

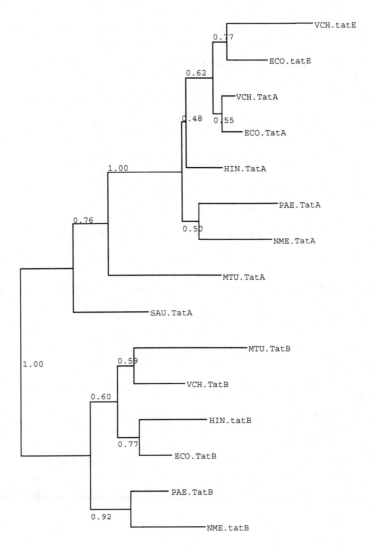

Fig. 3. The Tat tree with the rates of elementary quadruples

Always using the NJ method, we get the tree drawn in Fig. 3. The rate values of elementary quadruples are indicated along the edges (with the NJPlot

Table 4. Values of the criteria for the Tat tree of Fig. 3

Average length of external edges	0.33
Average length of internal edges	0.19
Ratio	$\rho = 0.56$
Criteria	
Distortion (in %)	7.7
Stress	0.105
Variance accounted for	.93
Rate of well designed quadruples	0.89
Rate of elementary quadruples	0.73
Arboricity	0.78

program (Perrière and Gouy, 1996). There are two edges reaching the greatest value 1; the first one separates clusters TatA and TatE from TatB and the other one, within the first cluster, separates bacteria Gram+ from Gram-. The criteria values, calculated with the Qualitree program are given in Table 4.

We have calculated the expected values, using the Modist program with $\rho = 0.56$, since it is the ratio of this tree. The result are very similar to those corresponding to $\rho = 0.50$, and we will use Table 1. According to the criteria, the Tat distance can be considered as a noisy tree distance with a rate equal to 20% ; $Vaf = .93$ indicate that $\tau = 15\%$. All the values belong to the confidence intervals. Consulting Table 2, and only the topological criteria, it can also be viewed as an euclidean distance in dimension 2, but metric criteria never fit whatever the dimension is. And there is no criterion except R_e having a value compatible with boolean distances. Consequently, we will accept the tree representation of the Tat distance and all the edges with a strong rate of elementary quadruples.

References

1. Bartélemy J.P. and Guenoche A.: Trees and Proximity Representations, J. Wiley (1991)
2. Gascuel 0. and Jean-Marie A.: Classification par arbre: la forme de l'arbre inféré dépend du schema algorithmique retenu, Actes des Septièmes Journées de la Société Francophone de Classification, SFC'99 (1999) 339-343
3. Guénoche A. and Préa P.: Counting and selecting at random phylogenetic topologies to compare reconstruction methods, Proceedings of the Conference of the International Federation of the Classifications Societies, IFCS'98, Short papers volume (1998), 242-245
4. Harding E.F.: The probabilities of rooted-tree shapes generated by random bifurcations, Advances in Applied Probabilities, 3 (1971) 44-77
5. Perrière and Gouy M.: WWW-Query: An on-line retrieval system for biological sequence banks. Biochimie, 78 (1996) 364-369
6. Pruzansky S., Tversky A. and Carroll J.D.: Spatial versus tree representations of proximity data, Psychometrika, 47, 1 (1982) 3-19
7. Saitou N. and Nei M.: The neighbor-joining method: a new method for reconstructing phylogenetic trees, Mol. Biol. Evol., 4 (1987) 406-425

8. Studier J.A. and Keppler K.J.: A note on the neighbor-joining method of Saitou and Nei, Mol. Biol. Evol., 5 (1988) 729-731.

9. Wu L.F., Ize B., Chanal A., Quentin Y. and Fichant G.: Bacterial Twin-arginine signal peptide-dependant protein translocation pathway: evolution and mechanism, J. Molecular Microbiology and Biotechnology (2000) (in press)

10. Wu L.F., Ize B., Chanal A., Quentin Y. and Fichant G.: Bacterial Twin-arginine signal peptide-dependant protein translocation pathway: evolution and mechanism, J. Molecular Microbiology and Biotechnology (2000) (in press)

Bayesian Approach to DNA Segmentation into Regions with Different Average Nucleotide Composition

Vsevolod Makeev[1], Vasily Ramensky[1], Mikhail Gelfand[2], Mikhail Roytberg[3], and Vladimir Tumanyan[1]

[1]Engelhardt Institute of Molecular Biology, Moscow, 117984, Russia
{makeev,ramensky,tuman}@imb.ac.ru
[2]VNIIGENETIKA, Moscow, Russia
mgelfand@ntl.ru
[3]Institute of Mathematical Problems of Biology, Puschino, Moscow Region, Russia
roytberg@impb.psn.ru

Abstract. We present a new method of segmentation of nucleotide sequences into regions with different average composition. The sequence is modelled as a series of segments; within each segment the sequence is considered as a random sequence of independent and identically distributed variables. The partition algorithm includes two stages. In the first stage the optimal partition is found, which maximises the overall product of marginal likelihoods calculated for each segment. To prevent segmentation into short segments, the border insertion penalty may be introduced. In the next stage segments with close compositions are merged. Filtration is performed with the help of partition function calculated for all possible subsets of boundaries that belong to the optimal partition. The long sequences can be segmented by dividing sequences and segmenting those parts separately. The contextual effects of repeats, genes and other genomic elements are readily visualised.

1   Introduction

1.1   Biological Motivation

Local nucleotide composition is believed important for many biological issues [1], [2] such as the isochoric organisation of the genome of higher eukaryotes [3], [4] compositional differences between exons and introns [5], [6], simple repeats (e.g. [7], tracts in splice sites [8] and binding sites [9] of DNA, GC islands in promoter sequences [10] and many others. Moreover, local nucleotide composition is accounted for in many algorithms developed to search different patterns in DNA sequences [11]. Usually the fixed length window is used and the results may undesirably depend on the length of the window.

O. Gascuel, M.-F. Sagot (Eds.): JOBIM 2000, LNCS 2066, pp. 57–73, 2001.
© Springer-Verlag Berlin Heidelberg 2001

1.2   Current Algorithms for Compositional Segmentation

Basically our approach is similar to that of [12], see also [13], [14]. The main difference is that Liu and Lawrence use weights for configurations with different number of segments, favoring segmentations to longer segments. Instead, we use the two-stage procedure with filtration of boundaries, which allows us to study segments with the chosen length-scale. By refusing to use weights we avoid an approximate procedure of sampling and can profit from the faster implementation of dynamic programming technique (N^2 instead of N^3, where N is the overall length of the sequence).

Another approach is developed in [15]. It uses the traditional frequency count estimator to which the Bayesian estimators converge for large segments. This approach is less justified for small segments. Recently several algorithms appeared employing hidden Markov models to obtain the segmentation of nucleotide sequences into segments with different composition [16], [17], [18]. However, in these models the number of possible compositional states usually is set *a priori*. This assumption works well when some particular DNA segments are searched for (for instance GpC islands [18]. At the same time this approach is less justified for partition of new genomic sequences, with no special attention paid to any particular region.

2   Optimal Segmentation

2.1   Probabilistic Formulation

A symbolic sequence over an alphabet W of L letters is considered as a series of segments, each segment is a Bernoulli type random sequence. Each segment has the corresponding symbol counts vector $\mathbf{n} = (n_1, \ldots n_L)$, where n_j is a number of occurrences of the j-th symbol in the segment; $\mathbf{n}$ has the multinomial distribution.

The Bayesian approach we are using [16], [19] regards the estimated parameters as random variables. In the beginning these variables have some prior probability distribution, which may be chosen rather arbitrarily. These probability distributions are re-estimated from the data using the Bayes formula, and the posterior distribution is obtained (see formula (4) below). The results of Bayesian estimation are always some probability distributions of the estimated quantity. Bayesian and classical statistics, however, agree for large samples because Bayesian distributions converge to the maximal likelihood estimation for any reasonable prior [20].

2.2   Choosing the Prior

We addressed the issue of choosing the appropriate prior in detail in [21]. The success of the Bayesian techniques depends dramatically on the prior used, especially for small samples, and the proper choice of the prior should be conditioned by the context of the problem, as there are no formal recipes.

For the segmentation problem, the choice of the prior actually implies some ideas about the overall composition of polymer. The simplest choice is the Dirichlet prior [12], [17], which reflects *a priori* information on the sequence composition. A more complex is the Dirichlet mixture prior [11], [17], which is based on the idea that a segment composition can come from one of several compositional classes. One can also use the entropic prior [19] reflects the statistical homogeneity of the prior source data.

The Dirichlet prior allows pseudocount interpretation [12], according to which the additional "pseudocounts" are added to the observed counts for each segment. Thus such the prior can be included into the algorithm without substantial modification of the formulae. The Dirichlet prior introduces some *a priori* composition of the sequence, and sequence segments, the composition of which is in contrast with this *a priori* composition, are more likely extracted at the same significance level. Thus, by choosing the proper prior one can adjust the program to extraction of specific functional region with known composition.

However, we believe that the noninformative prior we use, is more suitable for the initial segmentation of newly sequenced genomes. In this case the problem is not searching for specific regions, but assessing the general structure of the sequence. In practice, to choose the prior some statistical observations are made on data banks, or on the composition of the same sequence averaged on a larger scale. Thus, some correlation of the sequence composition with other sequences or with other parts of the studied sequence is introduced into the statistical inference, which is not always desirable, when the sequence under study is entirely new. Informative priors are better fit to the problem of searching for specific regions with at least approximately known composition, such as codons, GC islands etc.

2.3   Segment Likelihood

Denote the set of letter probabilities (the segment composition) as $\sigma = (\theta_1,\ldots,\theta_L)$; it complies to the normalisation condition:

$$\sum_{k=1}^{L} \theta_k = 1 \tag{1}$$

The likelihood of the individual sequence to occur is

$$L(\sigma) = \prod_{k=1}^{L} \theta_k^{n_k} \tag{2}$$

Given the composition $\sigma = (\theta_1,\ldots,\theta_L)$, write the probability density function $p(\sigma)$, which is defined on the simplex $S = \{\sigma : \theta_k \geq 0; \sum_{k=1}^{L} \theta_k = 1\}$. This probability density should comply with the normalisation condition

$$\int_S d\sigma p(\sigma) = 1 \tag{3}$$

Given some prior distribution $p(\sigma)$, consider the tentative block with counts $\mathbf{n}$. The Bayes theorem brings about the estimated probability density function $p(\sigma|\mathbf{n})$:

$$p(\sigma|\mathbf{n}) = \frac{L(\mathbf{n}|\sigma)p(\sigma)}{P(\mathbf{n})} \tag{4}$$

where

$$P(\mathbf{n}) = \int_S d\sigma L(\mathbf{n}|\sigma)p(\sigma) \tag{5}$$

is the normalisation constant called the marginal likelihood [12].

Marginal likelihood reflects the overall probability of obtaining the given sequence in the two stage random process. First, the composition σ is picked up according to the prior distribution, and then the sequence is generated in the Bernoulli random process with the letter probabilities σ. If $p(\sigma)$ is the uniform distribution on the surface of the simplex S, then

$$P(\mathbf{n}) = \frac{(L-1)!}{(N+L-1)!}n_1!...n_L! \tag{6}$$

Surprisingly, this quantity is also obtained in the conceptually similar but different probabilistic model [22]. For the sequence with the length N, the overall numbers of each letter $(n_1,...,n_L; \Sigma N_i=N)$ are picked up from the uniform in this case discrete distribution, then the probability of obtaining the sequence in the shuffling procedure is calculated. Since we consider the segments as independent, the complete likelihood of the sequence segmentation into K segments with known boundary location writes

$$P = \prod_{k=1}^{K} P_k(\mathbf{n}_k) \tag{7}$$

This quantity is optimised over the set of all possible boundary configurations yielding the optimal segmentation.

2.4   Dynamic Programming

The maximisation algorithm is formulated as follows. Consider a sequence $S = s_1s_2s_3...\,s_N$ of length N, where $s_i \in \Omega$. For every segment $S(a,b) = s_a...\,s_b$, $a \leq b$ with the length N, we introduce the weight $W(a,b)$. In our case $W(a,b) = \ln P(S(a,b))$. Any particular segmentation R in m blocks is determined by a set of boundaries $R = \{\ k_0=$

$0, k_1, \ldots, k_{m-1}, k_m = N$ }, where k_i separates s_{k_i} and s_{k_i+1}; $k_0 < k_1 < \ldots < k_{m-1} < k_m$. Define the weight of segmentation R as

$$F(R) = \sum_{j=1}^{m} W\left(k_{j-1} + 1, k_j\right) \qquad (8)$$

For functions determined on the segmentations, we shall also use another set of variables, the indicators of the boundary positions q_k, $k = 1, \ldots, N$. By definition, $q_k = 1$ if there exists a segment boundary after the letter k, otherwise zero. We shall use both variables, $F(R)$ and $F(q_1, \ldots, q_k)$, without special comments.

We are looking for the segmentation R^* that has the maximal weight. This is done in the recurrent manner. Denote by $R^*(k)$ the optimal segmentation of the fragment $S(1,k)$, $1 \le k \le N$. It is trivial to find $R^*(1)$. In the case of known optimal segmentations $R^*(1), \ldots, R^*(k-1)$, the optimal segmentation $R^*(k)$ is found using the following recurrent expression

$$F\left(R^*(k)\right) = \max_{i=0,\ldots,k-1}\left[F\left(R^*(i)\right) + W(i+1,k)\right] \qquad (9)$$

Here, we put $F(R^*(0)) = 0$. The recurrent relation (9) yields the algorithm. Since the building of segmentation $R^*(k)$ takes the time $\sim k$, the total time can be estimated as N^2.

2.5 An Example

Figure 1 displays the segmentation of a 1000 bp long random sequence with uniform composition (all letter probabilities are equal to 0.25). One can see that usually it is segmented into very short segments. Surely this is not what we would like to obtain. Practically, the segments containing many identical letters are extracted, but the general homogeneity of the block is never exhibited.

It should be noted that if we used a different prior, we would obtain a different segmentation pattern. For instance, for the prior, which corresponds to the composition rich with A and T, borders separating segments with different numbers of 'a' and 't' (such as a|tttt in the first line) are likely to disappear. Conversely, new boundaries can appear, as those separating segments with 'g' and 'c' (such as in the tgtt segment in the first line).

3. Border Insertion Penalties

3.1 Fluctuations in Local Composition

One can see (Fig. 1) that usually the segments of optimal segmentation are very short. Moreover, the random uniform Bernoulli sequence is also divided into many segments. When the sequence consists of several random homogeneous domains, the optimal segmentation includes many borders that are located within the domains.

Redundant boundaries are present due to statistical fluctuation in local composition in random sequences.

```
aaga| t | c| t gt t | aaca| g| a| t t t t | g| aa| t | g| c| a| t t t | g| a| c| g| cac| t ggt gt | c| t
| aa| g| t t | g| cc| t at | c| aa| c| t | gcg| t t | accaca| t t t | g| c| aga| c| gag| t t | c| a| c
gc| aa| g| t | c| g| t t | a| gg| t gt t t | gag| c| a| ct c| aa| c| g| t at t | c| gaaga| t t | c| g|
t ct | a| cgcg| t | g| c| aaa| cct cc| t | a| c| t | g| aa| t t | ggg| cc| g| t at at | gg| cc| g| a
a| c| t | g| t ct c| g| aca| t | g| t ct t ct t c| aaa| ccc| gagag| t | cc| aa| t | cc| t | accaca
cacc| g| aa| t gt t | a| g| cct c| gag| t | aa| c| agaga| c| t at aa| gg| t t | c| a| gg| t t t | c
c| g| t ct t | a| g| c| aa| ggg| t t t gt | c| t | gag| t | cccgc| g| t t | ggg| c| t | cgccg| aat a
| ccc| a| t | g| ct c| a| t | g| at a| ggagg| cc| ggg| c| t | a| gt gt t t g| c| a| gt gt | aa| t | c
| g| a| gt g| c| a| t t | c| a| g| t | gcgg| cc| t t | a| cc| aa| g| t t | g| cc| t gt t t | a| c| aaat
aa| gt g| aa| t | cc| gg| a| gcgcg| t t | c| g| t t | g| c| gag| c| t | cggcgc| t | gg| caacac|
gg| aa| g| t t | g| aaa| g| c| a| g| t t t | aga| t t | ccc| t t t | cc| g| t at aa| gg| at aa| t t gt |
ccgcg| t at a| g| t ct c| a| g| t | cgcc| gt t gt | a| cgc| t t | c| a| g| ccc| t | a| c| t | a| c| g
g| t at | cc| aga| cgc| a| t t | g| ccct cc| g| t | ccgcg| at a| ggcgg| t ct t t ct | a| t | c| g|
a| t ct | g| c| a| gg| t | a| cgc| gggt gg| aaca| g| t t | aa| c| t | aa| t cccct ct c| a| c| t t |
aa| t t | aa| cggcgc| a| gg| aa| cc| g| aaaca| c| g| t t | ccgcgc| t gt t | c| aa| t t | g| a| t
| aga| ccc| t | aa| gg| t t | aga| t t | c| at a| gg| at aa| t t | gagaa| cgc| t t t t at at | acaa
| g| t t | g| aa| c| gg| a| t | ggggg| t t | aa| gg| c| t | gg| c| a| t t | g| cccgc| t t t ct t ct | a
| t | cgc| t | a| g| c| t t | aa| c| g| aa| ggg| a| t | ccac| t | gcgagcg| ccc| at a| gg| t | a| g
| aacaa| g| cc| ggg| aaa| g| c| gagag| c| t | a| c| t t | g| a| t t t t | c| a| t | ccc| t | a| c| g
| t t | g| c| aaaa| cgc| t | aaat a| t t t t | gcg| at a| g| t t t t t gt t t | aca| t | gggt ggg| c| a
| t t t ct | gg| c| t t | caca| g| at a| ggg| t ct | g| c| t | a| g| t t | ccccacccc| t | g| cac| at
t at a| cac| g| t | a| t gt | c| a| t | g| a| t | g| ccgc| t | a| gg| aa| cct cc
```

Fig. 1. The optimal segmentation of a random sequence. Segment boundaries are marked with "|".

Thus it is advantageous to separate boundaries, which separate long regions with different compositions from those that reflect statistical fluctuations. This can be done by penalizing those segmentations that contains more boundaries. The penalty β for insertion a new border can be readily incorporated into functional (7)

$$ P = \sum_{k=1}^{K} P_k(\mathbf{n}_k)\beta^K \tag{10} $$

For $\beta < 1$ segmentations that contains less boundaries would have a preference. In our programs we included parameter $B = -\ln\beta$. The optimal value of B is chosen from computer simulations.

3.2   Random Sequences

It should be noted that the procedure we are using to estimate B is very heuristic. As it was pointed out by the referee of this paper, this parameter can be interpreted like a transition probability to enter in a new region. Then the problem could be presented as identical to modeling the sequence with a Hidden Markov Model (HMM) where the hidden states are the regions with homogeneous composition. This allows one to use one of many algorithms to estimate the hidden state transition in a HMM.

Thus, the main objective of the tests presented below is demonstration, that the border insertion penalty is the powerful tool in extracting homogeneous segments with statistically different composition in the sequence. We demonstrate, that the dependence of the optimal B values on the segment length and composition is very weak, thus the same B values can be employed for segmenting of sequences of different biological origin. Good statistical tools would help to increase the model performance, but the relevance of the model is reliably demonstrated by our empirical calculations. In segmenting a random sequence, one would like to obtain the result which complies to certain requirements. The first of those implies that a homogeneous random sequence should be segmented as a single block. We performed several series of statistical tests with sequences of different length and composition. We tested sequences of 100, 1000, and 10,000 symbols with different compositional biases. For each length and composition a hundred of random sequences was generated, for each of which the minimal B providing segmentation into the single unit was found. For the sequences of the same length, a greater bias in composition usually implied a smaller critical B value. For strongly biased sequences there were observed examples of segmentation into the single unit with $B = 0$. Greater B values are usually required to remove the inner boundaries from the longer random sequences. The histograms of the critical B values for the sequences with uniform (all $p_i = 0.25$) composition for different lengths are shown in Fig. 2. The dependence of the critical B value upon both the sequence length and the compositional bias is remarkably weak. As a rule, B of about 3–5 is enough to provide segmentation of a random sequence into the single block (compare with Figs. 3, and 4 in the next section).

4 Block Random Sequences

When the B value is taken too large the boundaries between longer regions with a homogeneous composition are also removed. To limit B from the above we performed two series of tests. In the first series of tests sequences consisting of two random blocks were generated. The B value for which the single inner boundary was present in the maximum number of block-random sequences were found. Our calculations demonstrated that for the given difference between semi-blocks for small lengths the adequate segmentation is never found (for any B). With the increase in the length of the semi-block the adequate segmentation is found for some share of the random sequences. The number of such sequences increases rapidly with the length of the random semi-blocks. This can be explained from statistical reasons: larger sequences mean richer samplings for estimations. The best B value, for which the two-part segmentation is obtained, can be estimated by equalizing overestimation and underestimation errors.

Here, overestimation corresponds to the case when for the chosen B the sequence is segmented into the single block, whereas the same sequence is correctly segmented in two blocks for a smaller B; respectively, underestimation corresponds to the case when for the same B the sequence is segmented into more than two segments, and the adequate segmentation is achieved for some greater B value.

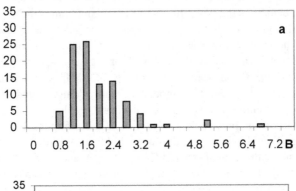

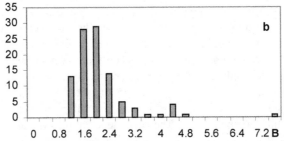

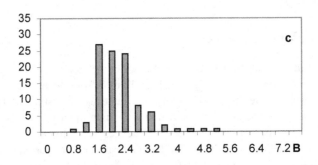

Fig. 2. Histogram of the minimal *B* necessary to segment a random homogeneous sequence into a single domain. 100 experiments. The uniform probability distribution. The sequence length is (*a*) 100bp; (b) 500 bp; (c) 1000bp

The optimal B value for 100 letter blocks with probabilities of (0.2, 0.2, 0.3, 0.3) and (0.3, 0.3, 0.2, 0.2) was about 2.7. From 100 random sequences tested, 60 were adequately segmented in two parts. Other 40 were not segmented into two parts for any B. The optimal value B = 2.7 was underestimation for six sequences and overestimation for seven ones. Thus, 45 sequences, a little less than a half, were adequately segmented. For B = 5, only two sequences were segmented in three blocks (both contained several identical letters on one of the ends), 25 of the sequences were segmented adequately with one boundary in the middle, and 75% sequences were segmented as a single unit.

However, if the bottom limit of B increases very slowly with the length, the upper limit increases dramatically faster. For instance, all the sequences made from two 1000–letter random blocks with the same probabilities as in the previous example are segmented in two blocks for any B from the range from 7.2 to 23.5 with the bottom limit for B less than 5 for 96 examples and less than 3 for 92 examples out of 100. This allows one to choose B rather comfortably when the short domains are not very important.

Obviously, the longer the segments, the smaller the difference, which the method is possible to resolve. The sequence made of two 1000 blocks with compositions (0.24, 0.24, 0.26, 0.26) and (0.26, 0.26, 0.24, 0.24) is almost never (in <10% of all cases) segmented adequately. However, the single boundary in the sequence with the same (0.24 vs. 0.26) block probabilities can be reliably resolved for the segment length of 10,000. The error for the boundary location also grows with the decrease in the compositional difference.

Similar results were obtained in the experiments on island recognition (Figs. 3 and 4). Random sequences were generated and in each experiment islands of contrasting compositions were put at random into the uniform sequence with the length of 1000, 2000, and 5000 length. The island length amounted to 0.1 of the total sequence length. We scanned over all BIPs with the 0.5 step. We put that the island was recognized if the strictly two inner boundaries were found in the sequence allowing for an error of 10% of island length. The example of the dependence on the BIP is shown in Fig. 3.

One can see, that if the compositional contrast is significant enough, to provide an acceptable recognition of the segment, the B parameter can be picked up rather comfortably. Fig. 4 shows the example of trade off between the compositional difference and the length of the segment.

In the next series of tests we generated sequences consisting of 10 blocks with different compositions and determined the B values for which the correct number of blocks was obtained. Then, we monitored the positions of boundaries in the segmentation with this optimal B. For the 100–letter blocks some false-positive boundaries (often near the ends of the sequence) were added, and some blocks with close compositions were merged. The chance of block merging is much greater for the blocks whose composition is close to the uniform ($p_i = 0.25$ for all is). Again the percentage of correctly segmented sequences was greater for the sequences consisting of the longer blocks. In all examples of 10,000–letter blocks segmented with $B = 3.5$ we obtained the adequate segmentation.

All in all, it seems that choosing B between 3 and 5 allows one to eliminate most of fluctuations and consider the domains obtained as random with a fixed composition. For $B \gg 5$, almost only the adequate boundaries are left, but some presumably divergent blocks are merged.

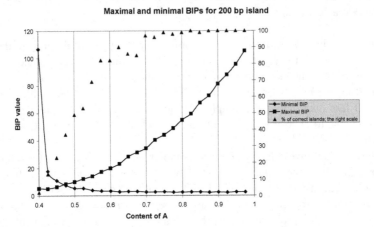

Fig. 3. Searching for an island of contrasting composition within a long sequence. The island with the length 200bp is located at the random place in the random sequence of the uniform composition with the length 2000 bp. Along the X-axis is the A-content of the island, the other three letters have the identical content. For each composition scanning by the BIP value is performed from bottom to top. The minimal BIP is the average BIP (averaged over 100 experiments) at which the island is correctly recognised (if any). The maximal BIP is the average BIP corresponding to the limiting value, above which the sequence is segmented into the single segment. The right scale shows the percentage of sequences for which the island was correctly recognised allowing for 20bp error in the boundary positions.

5   Filtration of Boundaries

5.1   Partition Function

Another way to study the relative significance of a boundary is the partition function. With the help of this function one can calculate a score, which reflects how the addition of this particular boundary influences weights of segmentations.

Given the probability of each segmentation, the partition function of the segmentations can be determined a standard way [23] by summing the probabilities of all possible partitions:

$$Z(N) = \sum_{q_1} \ldots \sum_{q_{N-1}} \Pi(q_1, \ldots q_{N-1}) \tag{11}$$

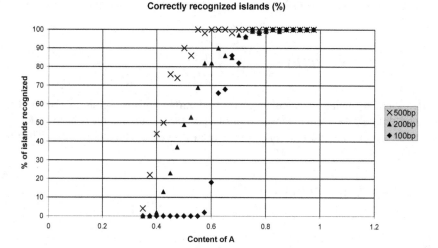

Fig. 4 Dependence of the percentage of correctly recognized "islands" on the island" composition and the island length. The islands of the lengths of 100, 200, and 500 bp were placed at random into the sequence with the uniform composition, the length of the islands was always 0.1 of the total sequence length. Along the X-axis is the A-content of the island, the other three letters have the identical content. $B = 5$ in all experiments.

here q_k equals unity if there is a segment boundary after the letter k in the sequence, zero otherwise; the overall $\mathbf{q} = (q_1,...,q_N)$ determines a segmentation which has the probability $\Pi(\mathbf{q})$.

With the partition function at hand we can calculate the probability of a boundary to be located after a particular letter k, via calculating two partition functions of the regions to the left and to the right of this border, ZL and ZR respectively:

$$\Pi(k) = \frac{ZL(k)ZR(N-k)}{Z(N)} \qquad (12)$$

5.2 Dynamic Programming

The partition function (11) may be rewritten as follows [23]:

$$Z(N) = \sum_{q_1} ... \sum_{q_{N-1}} e^{F(q_1...q_{N-1})} \qquad (13)$$

To calculate the probability of a boundary after the letter k, we need also the partition functions of the segments to the left and to the right of this boundary:

$$ZL(k) = \quad \ldots \quad e^{F(q_1, \ldots, q_{k-1})} \tag{14}$$
$$\quad\quad q_1 \quad\quad q_{k-1}$$

$$ZR(k) = \quad \ldots \quad e^{F(q_k, \ldots, q_{N-1})} \tag{15}$$
$$\quad\quad q_k \quad\quad q_{N-1}$$

Recurrent formulae to calculate $ZL(k)$ and $ZR(k)$ are analogous to (10) and are obtained through the formal substitution of operations. Summation is used instead of taking the maximum, and multiplication is used instead of summation [23]. Then the following relations replace (10):

$$ZL(k) = \sum_{j=0}^{k-1} e^{W(j+1,k-1)} ZL(j) \tag{16}$$

$$ZR(k) = \sum_{j=k}^{N} e^{W(k,j)} ZR(j) \tag{17}$$

with the respective boundary conditions $ZL(0) = ZR(N+1) = 1$; $W(k-1, k) = W(N, N+1) = 0$.

The obvious modification of the dynamic programming calculates the partition function in the case when only the given set of boundaries is allowed.

5.3   Filtration Strategy

For the best result one should combine calculation of optimal segmentation with filtration. At the first stage the optimal segmentation with some B is found. Then the cutoff value is chosen and all the boundaries with probabilities (11) lower than that cutoff value are removed. The resulting set of boundaries usually is not optimal in the sense that some boundaries from the resulting set are removed when the step one is repeated. So an additional round of optimisation (with the same B) is performed, removing some boundaries. Iterations converge rapidly to the stable set of boundaries all of which have the partition function probabilities greater than the cutoff value.

5.4   Partition Function Cutoff vs. Border Insertion Penalties

Both partition function cutoffs and border insertion penalties serve for the same purpose: to merge segments with close composition absorbing local fluctuation of composition. However, two segmentations into the same number of domains obtained via these two methods are not the same. Segmentation with given B is the best segmentation from all the segmentations with the fixed number of domains (this is clearly seen by taking log in (7)). Thus if one obtains two segmentations with the same number of domains, one using B and the other using some filtration cutoff level, then the segmentation obtained via border insertion penalty would have a greater score. However, the difference is not the critical and in general two segmentations agree.

5.5   An Example of the Large Scale Segmentation

We have segmented first 200,000 bp of the complete sequence of *Plasmodium falciparum* chromosome II available at www.tigr.org. This sequence contains a telomeric region, a long subtelomeric repeat and several genes, which consist of exons with various length. This genome is rich with A+T (in average 80%). In contrast because a gene should code for the protein its AT content cannot be to high, since some codons contain G and C in the significant positions of the triplet. Thus one can hope that some long segments in this sequence are related to the coding sequences.

To evaluate this hypothesis, 200,000 bp sequence was segmented with $B = 3$. Filtration with 0.999 cutoff level was performed. All the segments longer than 500 with the G+C content greater than 0.2 were taken.

From the chromosome description all exons with the length greater than 200 were taken. The results of the comparison are shown in the Table. The telomeric repeat and the long 21999 bp subtelomeric repeat found in this sequence is clearly seen. Among remaining 41 long GC containing segments 30 coincide with long exons with different precision (marked with "y"). In one case (177,844) the segment contains two exons. In one case (149,594) three subsequent segments cover one long exon. Seven long exons hadn't any corresponding segments. These are false negative of our "prediction".

There are five examples, when long GC containing segments did not contain long exons. We searched for long ORFs in such segments. Indeed, three segments out of five contained long ORFs (starting at 27,864; 112521; and 192,716). These are good candidates for the new-found genes. Two segments (at 23,281 and 171,499) did not contain long ORFs. These are false positives of our "prediction". One can see that our simple model of a long exon (long GC segment containing long ORF) describes surprisingly well the situation found in *Plasmodium falciparum* chromosome II.

The fact that the coding regions are more compositionally uniform than the uncoding regions has been reported by several authors, who used different segmentation procedures and different biological material. It was reported in [24] for compilations of coding and non-coding sequences from different organisms were studied. The HMM results published in [11] also allows one to make such conclusion. The attempt to solve an inverse problem, that is to find coding regions by the segmentation procedure was published in [25] for *Rickettsia prowazekii*. The authors of [25] used an entropic segmentation, which is similar to our approach in the case of long segments (see [21] for comparison), however, they encoded DNA sequence in 12 letter alphabet associated with the codon triplet preferences. Thus their algorithm is more related to the statistical method of gene finding and in a sense is related to the HMM method, published in [11].

Special methods of statistical gene finding may have a better performance as compared to our general segmentation method. Here we want to note, that in *Plasmodium falciparum* genome, the majority of long exons can be extracted *only* by compositional segmentation, which we believe implies that in reality there are not so many examples of homogeneous regions in genomes, besides coding regions and repeats. The preliminary data that we obtained for *Leishmania major* confirm this conclusion.

6   Discussion

When the initial segmentation with the non-informative prior obtained, it becomes interesting to cluster the composition of the resulting segments. Firstly, our preliminary calculations on *Plasmodium falciparum* and *Leishmania major* genomes demonstrated that there are limited number of classes of long, statistically homogeneous regions in genomes of lower eukaryotes. These are long exons, divergent repeats, and other low-complexity regions, such as AT-rich strands. Short exons and intergenic regions are segmented into many short segments.

Table 1. Comparison of long GC-rich segments with coding regions in initial 200,000bp segment of *P.falciparum*

SegBeg	SegEnd	SegLen	Score	GC	ExBeg	ExEnd	ExLen
0	1152	1152	-0.236931	0.4635	telomeric repeat		
1152	23151	21999	1392.596229	0.3237	long sub-telomeric repeat		
23281	24034	753	63.886702	0.2922			n
25106	27469	2363	139.753512	0.3275	25232	29035	3803 y
27861	29136	1275	70.190261	0.3318	27864	29183	1319 long ORF
29658	31160	1502	140.942212	0.2843	29837	31168	1331 y
32952	33956	1004	41.621573	0.3516	33030	33965	935 y
35868	37186	1318	120.276477	0.2868	35927	37249	1322 y
38300	39105	805	39.909483	0.3379	38287	39132	845 y
41439	42558	1119	46.975472	0.3512	41515	42573	1058 y
					45286	46344	1058 n
47186	49879	2693	499.468944	0.2042	48923	49861	938 y
51774	53139	1365	132.519112	0.2806	52456	53202	746 y
54351	55083	732	28.625235	0.3538	54418	54936	518 y
57445	58207	762	21.842848	0.3727	57344	58228	884 y
63358	64231	873	41.500332	0.3414	63360	64376	1016 y
66723	67531	808	41.432232	0.3354	66729	67545	816 y
					69370	69771	401 n
73492	74070	578	58.532224	0.2734	73441	74094	653 y
77158	78522	1364	160.958018	0.2595	77251	78360	1109 y
81349	83887	2538	317.825703	0.2537	81291	83900	2609 y
86838	87409	571	45.914259	0.296	86832	87400	568 y
					87675	88110	435 n

91330	98457	7127	568.747181	0.3022	91318	98532	7214	y
103364	105245	1881	57.568751	0.3732	103385	105238	1853	y
109570	110596	1026	141.488399	0.2407	109564	110202	638	y
112528	113189	661	76.080559	0.2602	112551	113167	616	long ORF
117633	118436	803	129.753598	0.2204	117558	118167	609	y
120595	124170	3575	401.15354	0.2666	120524	124102	3578	y
					127994	128314	320	n
129441	133827	4386	696.118369	0.2253	129688	133570	3882	y
					135523	137139	1616	n
					139955	140191	236	n
141536	147622	6086	1096.926023	0.2087	141625	147564	5939	y
149524	151241	1717	207.214863	0.2574	149524	156981	7457	y
151257	153666	2409	425.008465	0.2109	cont			y
153781	157075	3294	595.192248	0.208	cont			y
158090	159707	1617	277.641574	0.214	158137	159660	1523	y
160425	161310	885	138.391867	0.2249	160514	161242	728	y
					166144	168051	1907	n
168815	170150	1335	105.964553	0.3004	168838	170136	1298	y
171499	172084	585	82.778237	0.2359				n
177123	178030	907	86.610189	0.2811	176628	178300	1672	y
178844	181560	2716	470.085051	0.2135	178955	180924	1969	y
cont					181103	181526	423	
183428	185860	2432	228.672468	0.285	182228	189115	6887	y
192622	194059	1437	271.130882	0.2011	192716	193324	608	long ORF
194795	198041	3246	597.246378	0.2055	194826	196916	2090	y
198302	199601	1299	162.87607	0.2525	198353	199570	1217	y

Although these results are preliminary and should be tested on a greater number of genomes, it is very likely that segment compositions in native sequences belong to the limited number of classes. In this case the advanced segmentation algorithm, for instance Hidden Markov Models, which uses the number of compositional states as the parameter becomes entirely relevant [18]. Knowing natural compositional classes will facilitate constructing a good informative prior, which allows one to reliably annotate genomic sequences with fast segmentation method.

Moreover, the comparative positioning of the regions with distinct composition can be an interesting for assessing evolution, one of whose basic processes is the relocation of parts of genomes between chromosomes or within the same chromosome [26].

Thus, the segmentation can yield a lot of valuable biological information. We believe that our method suits best for the initial segmentation of newly sequenced genomes, with no *a priori* information on the composition. Another possible application is initial segmentation into long regions (with large B) as a preprocessing before pattern search procedure, which often uses general composition as a statistical reference. The power of such procedure can be increased by referring the algorithms not to the general composition of the sequence but to the composition of the region.

Compositional dependences can be helpful in searching for many functionally important patterns. However, in this case a more specialized algorithms appear to be more powerful. We believe that HMM is the best for such the case. Our approach is faster than HMM, since we use no sampling procedure, which requires thoughtful convergence control [18]. Thus, our algorithm can be helpful in studies of long complete genomes. In this case the application of informative prior constructed with the reference to the specific region can improve the results.

Acknowledgements. The authors wish to thank D. Haussler for stimulating discussion and for the reference to the work of Liu and Lawrence (1998), R. Guigo for discussion and for the copy of the proceedings of the Halkidiki (1997) workshop (Lawrence, 1997), D. Cellier for the inspiriting discussion and G. Shaeffer and M. Regnier for many valuable comments. We also wish to thank the anonymous referee for many valuable remarks. This study has been partially supported by Russian Human Genome Program 18/48 (V.E.R., V.Ju.M., V.G.T), and 99-0153-F(018) (M.A.R.), INTAS grant 99-1476, Lyapunov Institute Project, and MGRI project 244.

References

1. Karlin, S., Brendel, V.: Patchiness and correlation in DNA sequences. Science **259** (1993) 677–680.
2. Li, W.: The study of correlation structure of DNA sequences: a critical review. Computer & Chemistry **21(4)** (1997) 257–278.
3. Bernardi, G.: The isochore organization of the human genome. Annual Review of Genetics **23** (1989) 637–661.
4. D'Onofrio, G., Mouchiroud, D., Aissani, B., Gautier, C., Bernardi, G.: Correlation between the compositional properties of human genes, codon usage, and amino acid composition of proteins. J. Mol. Evol. **32** (1991) 504–510.
5. Guigo, R. Fickett, J. W.: Distinctive sequence features in protein coding, genic noncoding and intergenic human DNA. J. Mol. Biol. **253** (1995) 51–60.
6. Herzel, H., Grosse, I.: Correlation in DNA sequences: The role of protein coding segments. Phys. Rev. E. **55** (1997) 800–810.
7. Li, W., Kaneko, V.: DNA Correlations. Nature **360** (1992) 635–636.
8. Gelfand, M. S.: Prediction of function in DNA sequence analysis. Journal of Computational Biology **2** (1995) 87–117.
9. Gelfand, M. S., Koonin, E. V.: Avoidance of palindromic words in bacterial and archaeal genomes: a close connection with restriction enzymes. Nucl. Acid. Res **27** (1995) 2430–2439.
10. Pedersen, A. G., Baldi, P., Chauvin, Y. Brunak, S.: The biology of eukaryotic promoter prediction. Computer & Chemistry **23** (1999) 191–207.
11. Krogh, A., Mian, I. S. Haussler, D.: A hidden Markov model that finds genes in E.coli DNA. Nucl. Acid. Res **22** (1994) 4768–4778.
12. Liu, S. L., Lawrence, C. E.: Bayesian Inference of Biopolymer Models. Bioinformatics **15** (1999) 38–52.

13. Lawrence, C. E.: Bayesian Bioinformatics. 5th international conference on intelligent systems for molecular biology, Halkidiki, Greece (1997).
14. Liu, S. L., Lawrence, C. E.: Bayesian inference of biopolymer models, Stanford Statistical Department Technical Report (1998).
15. Roman-Roldan, R., Bernaola-Galvan, P. and Oliver, J. L.: Sequence compositional complexity of DNA through an entropic segmentation method. Phys. Rev. Lett. **80** (1998) 1344.
16. Churchill, G. A.: Stochastic models for heterogeneous DNA sequences. Bull. Math. Biol. **51** (1989) 79–94.
17. Durbin, R., Eddy, Y. S., Krogh, A. Mitchison, G.: Biological Sequence Analysis. Cambridge, Cambirdge University Press (1998).
18. Muri, F., Chauveau, D., Cellier, D.: Convergence assessment in latent variable models: DNA applications. In C. P. Robert (ed.) Lectural Notes in Statistics, Vol. 135, Discretization and MCMC convergence assessment., Springer. (1998) 127–146.
19. Wolpert, D. H., Wolf, D. R.: Estimating functions of probability distributions from a finite set of samples. Phys. Rev. E. **52** (1995) 6841–6854.
20. Rozanov, Y. M.: Teoriya veroyatnosti, sluchainye processy i matematicheskaya statistika (russ: Probability Theory, Stochastic Processes and Mathematical Statisitics). Moscow, Nauka (1985).
21. Ramensky, V.E., Makeev, V.Ju., Roytberg, M.A., Tumanyan, V.G.: DNA segmentation through the bayesian approach. Journal of Computational Biology., **7** (2000), 215–231.
22. Shaeffer, G. (1999) Personal communication.
23. Finkelstein, A. V., Roytberg, M. A.: Computation of biopolymers: A general approach to different problems. BioSystems **30** (1993) 1–19.
24. Ossadnik, S.M., Buldyrev, S.V., Goldberger, A.L., Havlin, S., Mantegna, R.N., Peng, C.-K., Simons, M., Stanley, H.E.: Correlation approach to identify coding regions in DNA sequences. Biophysical Journal **67** (1994) 64–70.
25. Bernaola-Galván, P., Grosse, I., Carpena, P., Oliver, J., Román-Roldán, R., Stanley, H.: Finding borders between coding and noncoding DNA regions by an entropic segmentation method. Phys. Rev. Let., **85,** (2000) 1342–1345.
26. Ono, S.: Evolution by gene duplication. Springer. (1970)

Exact and Asymptotic Distribution of the Local Score of One i.i.d. Random Sequence

Sabine Mercier[1], Dominique Cellier[2], François Charlot[2], and
Jean-Jacques Daudin[3]

[1] UFR SES, Département Mathématique et Informatique Université de Toulouse II,
5 allée Antonio Machado 31058 Toulouse cedex, France
mercier@univ-tlse2.fr
[2] Université de Rouen, Analyse et Modèles Stochastiques UPRES A 6085, 76821
Mont-Saint-Aignan cedex, France
{cellier,charlot}@univ-rouen.fr
[3] Institut National Agronomique Paris-Grignon, Département OMIP, UMR
INAPG-INRA 96021111, 16, rue Cl. Bernard, 75231 Paris cedex 05, France
daudin@inapg.inra.fr

Abstract. We propose two new and complementary methods to assess
the statistical significance of high scoring segments, within both long and
short sequences. The numerical results show that these methods improve
the work of Karlin *et al.* implemented in BLAST for the comparison of
two sequences.

Keywords: local score, statistical significance, sequence analysis.

1 Introduction

In order to highlight interesting segments of biological sequences, numerical values called scores, integer or rational, are assigned to the components of the sequences, the amino acids or nucleotides, reflecting their physical and chemical characteristics like hydrophoby, acidity, potential... See [5] or [7] for examples of scoring functions.

Let us take the acidic score function s given by Karlin *et al.* in their article [5]: $s = +2$ for D, E ; $s = -2$ for K, R ; and $s = -1$ for the others. To a given sequence $\mathbb{S}$, corresponds the score sequence $\mathbb{X}$ which associates a score with each component.

$$\mathbb{S}: V\ L\ S\ E\ A\ D\ K\ T\ V\ K\ A\ E\ W\ E$$
$$\mathbb{X}: \text{-1 -1 -1 +2 -1 +2 -2 -1 -1 -2 -1 +2 -1 +2}$$

The score of a segment $S_i...S_j$ is equal to the sum of the scores included in this segment

$$s(S_i...S_j) = \sum_{k=i}^{j} s(S_k) \ .$$

O. Gascuel, M.-F. Sagot (Eds.): JOBIM 2000, LNCS 2066, pp. 74–83, 2001.

We define the local score of a sequence of length n, and we note H_n, the maximum obtained considering every segment of the sequence:

$$H_n = \max_{1 \leq i \leq j \leq n} \sum_{k=i}^{j} s(S_k) = \max_{1 \leq i \leq j \leq n} \sum_{k=i}^{j} X_k \ ,$$

Due to the choice of scoring function, the segment that realizes the local score is the most acidic segment of the protein $\mathbb{S}$. In the previous example, there are two tied winning segments, each with a local score of 3. Therefore $H_{16} = 3$.

The question asked to the statistician is thus the following: could this result have been obtained by chance ? To assess the significance of the obtained score one must assume an underlying M0 model. The underlying model used in this work is the one generally used. Biological sequences are represented as sequences of identically and independently random variables. Therefore we have to compute the P-value under M0, that is $P[H_n \geq a]$, where H_n is the local score of the random sequence and a is the actual obtained local score, in order to determine the statistical significance of this result. Karlin *et al.* proposed an approximation for the P-value under the two following conditions: i) $E[X] < 0$ and ii) $P[X > 0] > 0$ (see [5] [6]).

$$\lim_{n \to +\infty} P[H_n \leq \frac{\log n}{\lambda} + x] = \exp(-K^* \cdot e^{-\lambda x}) \ , \tag{1}$$

where λ and K^* depend only on the probability distribution of X. This approximation is used in BLAST (see [1]) in the case of the alignment of two sequences (see [4] for the generalization of (1) to the alignment of two sequences).

In this paper, we give a more accurate approximation in Sect. 2 and the exact P-value in Sect. 3. These results are independent from each other and are obtained using different derivations, such as the one from Karlin *et al.*

2   New Asymptotic Distribution for the Local Score

Let $\mathbb{X} = (X_i)_{i \geq 1}$ be a sequence of independent and identically distributed random variables having values in $\mathbb{Z}$ (rational case can be deduced). We assume like Karlin *et al.* that $E[X] > 0$ and $P[X > 0] > 0$. Let M be the maximum of partial sums:

$$M = \max_{0 \leq k}(X_1 + ... + X_k) \ .$$

Karlin *et al.* used the renewal theory to get an approximation of the distribution of M, allowing them to get Formula (1).

Using the random walk theory, we establish the exact p.d.f. of M. This result allows us to obtain additional terms in the asymptotic expression of the p.d.f. of the local score H_n and brings a more accurate approximation.

Section 2.1 contains more details for the distribution of M. Some applications are developed in Sect. 2.2 and derived results on the distribution of the local score are given in Sect. 2.3.

Demonstrations concerning M and the local score are quite long. They are not explained in this article, but are done in details in [8].

2.1   Exact Distribution of M

Let W^x denote the Lindley process,

$$W_0^x = x \qquad \text{and} \qquad W_k^x = (W_{k-1}^x + X_k)^+ . \tag{2}$$

W^x is a classic Markov chain used in queue theory.

For $x = \sup_{p<0, p \in \mathbb{Z}}(\sum_{i=p+1,...,0} X_i)$, we show (see [8]) that M is the unique invariant probability measure, noted γ, of $(W_k^x)_{n \geq 0}$. Noting Λ the transition matrix of W^x, we have

$$\gamma \cdot \Lambda = \gamma . \tag{3}$$

The matrix Λ depends only on the distribution of X. The distribution of M is totally determined by relation (3) and the fact that $\sum_{k \geq 0} \gamma_k = 1$.

By developing (3), we get a linear system of different equations including the γ_k. The main one allows us to deduce that γ_k can be expressed as a linear combination of sequences defined by successive powers of roots, noted R_i, of a polynomial linked to the recurrent linear equation.

2.2   Properties of the Roots and Applications

As γ is a probability, we can deduce that only roots with a module $|R_i|$ smaller than 1 must be considered (where $|R_i| = (\mathcal{R}(R_i)^2 + \mathcal{I}(R_i)^2)^{1/2}$, $\mathcal{R}(R_i)$ the real part of R_i and $\mathcal{I}(R_i)$ its imaginary one).

The main relation extracted from (3) is such that the polynomial has only two real positive roots: 1 and the second one noted R, with $0 < R < 1$. We show (see [8]) that any other root R_i with $|R_i| < 1$ verifies $|R_i| < R$. So the term linked to R corresponds to the main term of the expression.

R and λ, defined in (1) by Karlin *et al.*, are linked as follows: $\lambda = -\log R$.

Application for High-scoring Mixed Charge Segments. Amphoteric (or mixed charge) amino acids can be positively or negatively charged depending on the environment they are in. For high-scoring mixed charge segments, we use the scoring scheme given by Karlin *et al.* in [5]: $s = 2$ for Aspartate (D), Glutamate (E), Lysine (K), Arginine (R) and l'Histidine (H); $s = -1$ for the others. Let $p = P[X = 2]$ be the probability of getting the five amino acids cited previously.

The polynomial derived from (3) is $P(x) = (1-p)x^3 - x^2 + p$. The two roots which are different from 1 re real.

$$0 < R = \frac{p + \sqrt{p(4 - 3p)}}{2(1-p)} < 1 , \tag{4}$$

and

$$-1 < -R' = \frac{p - \sqrt{p(4 - 3p)}}{2(1-p)} < 0 . \tag{5}$$

The distribution of M is given by

$$(\forall k \geq 0) \quad \gamma_k = P[M = k] = \tau \cdot \left[R^{k+1} + (-1)^k R'^{k+1} \right] , \tag{6}$$

with

$$\tau = \frac{(1 - 3p)}{\sqrt{p(4 - 3p)}} . \tag{7}$$

Application for High-scoring Acidic Charge Segments. The scoring function for the acidic case is given in Sect. 1. Let us denote $p_2 = P[X = 2]$, $p_{-1} = P[X = -1]$ and $p_{-2} = P[X = -2]$. We have $P(x) = p_2 x^4 + p_{-1} x^3 - x^2 + p_2 = (1 - x)Q(x)$, with $Q(x) = -p_{-2} x^3 - (1 - p_2)x^2 + p_2 x + p_2$. A simple study of Q shows that there are four real roots for P: 1, the main R, and R' and R'', where only R and R' belong to $] - 1; 1[$. We also have $R' < 0$ and $|R'| < R$. We get

$$(\forall k \geq 0) \quad \gamma_k = P[M = k] = \tau \cdot \left[R^{k+1} - R'^{k+1} \right] , \tag{8}$$

with

$$\tau = (1 - R)(1 - R')/(R + R') . \tag{9}$$

2.3   Local Score

The result on the distribution of M brings a new asymptotic formula for the p.d.f. of H_n. This distribution is built up from a linear combination of successive powers of the roots R_i as well. The general formula seems to be complicated but in every example we studied (the ones proposed in [5]), the formulas are simple, as there are only two roots of interest.

Result 1 (Asymptotic Distribution for Mixed Charge Case)
The distribution of the local score is approximated by the formula

$$P\left[H_n \leq \frac{\log n}{\lambda} + x \right] \stackrel{n \to +\infty}{\sim} \left[1 - \frac{\tau R^{x+2}}{n} + (-1)^{\beta_n(x)} \cdot \frac{\tau R'^{x+2}}{n^\alpha} \right]^{\frac{n}{\mu}+1} , \tag{10}$$

where $\lambda = -\log R$, $\beta_n(x) = \lfloor \frac{\log n}{\lambda} + x \rfloor$, $\alpha = \log R'/\log R$, $\mu = -1/E[X]$, and where R, R', τ are given in (4), (5) and (7).

With $K = \tau R^2$, we deduce from Result 1 that

$$P\left[H_n \leq \frac{\log n}{\lambda} + x \right] \stackrel{n \to +\infty}{\sim} \exp\left(-\frac{K}{\mu} R^x \right) .$$

It is not difficult in this application to verify that $K/\mu = K^*$, with K^* defined by Karlin *et al.*, see (1). So we can deduce from (10) Formula (1).

Result 2 (Asymptotic Distribution for Acidic Case)

$$P\left[H_n \leq \frac{\log n}{\lambda} + x\right] \overset{n\to+\infty}{\sim} \left[1 - \frac{\tau R^{x+2}}{n} + (-1)^{\beta_n(x)} \cdot \frac{\tau |R'|^{x+2}}{n^\alpha}\right]^{\frac{n}{\mu}+1} \tag{11}$$

where $\lambda = -\log R$, $\alpha = \log|R'|/\log R$, $\beta_n(x) = \lfloor \frac{\log n}{\lambda} + x \rfloor$, μ *depends on the parameters of the model,* R *and* R' *are the two roots of* P *(see Sect. 2.2, acidic application), and where* τ *is given in (9).*

Applications are implemented for the research of the most mixed-charged segments. We numerically test Formula (10) by comparing the two different approximations with an empirical distribution. The empirical distribution has been calculated on simulated sequences. The comparison shows the real improvement of the method (see Fig. 1).

For some parameters, this approximation is very accurate (see Fig. 1) even for sequences of length less than 500, but for others the approximation is of interest only for long sequences (see Fig. 2).

The short sequences problem is solved by the next approach.

3   Exact Distribution for the Local Score

We consider again $\mathbb{X} = (X_i)_{i\geq 1}$ a sequence of independent and identically distributed random variables having values in $\mathbb{Z}$. This section generalizes the hypothesis in light of the present work and that of Karlin *et al.*: the expected score $E[X]$ is allowed to be negative, positive or equal to zero. The method establishes the exact distribution of H_n and is based on current results on Markov chains. The local score can be expressed as the maximum of a Lindley process often used in Waiting Line theory (see (2) for definition of a Lindley process). We have

$$H_n = \max_{0\leq k\leq n} W_k^0 .$$

More details and precised proof can be found in [2] and [9].

Let us denote $f(k) = P[X \leq k]$ and $p(k) = P[X = k]$. Let Π be the following matrix, and a a positive integer. For $0 \leq h, \ell \leq a$,

$$\Pi = \left(\begin{array}{cccc|c}
f(0) & p(1) & & p(a-1) & 1-f(a-1) \\
\vdots & & \vdots & & \vdots \\
f(-h) & \cdots & p(\ell-h) & & 1-f(a-h-1) \\
\vdots & & & & \vdots \\
f(1-a) & p(2-a) & & p(0) & 1-f(0) \\
\hline
0 & 0 & \cdots & 0 & 1
\end{array}\right).$$

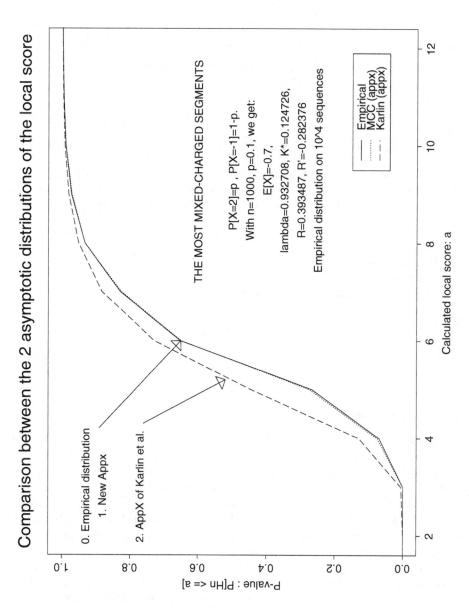

Fig. 1. The Most mixed-charged Segment. Here is an example of the real improvement of the new approximation even for not-so-long sequences, here $n=1000$. The curves of the empirical distribution and of the approximation we propose are nearly the same.

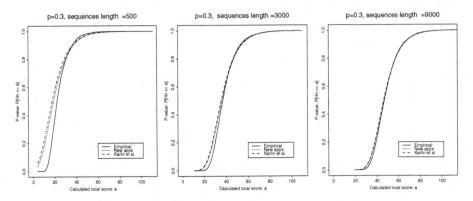

Fig. 2. The Most Mixed-charged Segment. For $p = P[X = 1] = 0.3$ the two approximations are not so good for short sequences. We can see the influence of the length of the sequences: the longer the sequences are, the better the asymptotic distributions are.

Π is the probability transition matrix of the Lindley process linked to the score sequence $\mathbb{X}$.

Thus the probability that the local score H_n reaches a is given by

Theorem 1.

$$P[H_n \geq a] = P_0 \cdot \Pi^n \cdot Q_0 , \qquad (12)$$

with P_0 the horizontal vector $(1,0,0,...,0)$ of dimension $a+1$ and Q_0 the vertical vector $(1,0,0,...,0)$ of dimension $a+1$.

3.1   Exact Distribution in Practice

To obtain $P[H_n \geq a]$ we have to compute the n-th power of a square matrix of size $(a+1)$. This is simple to program in few lines of code. A MATLAB computer program for this purpose (available upon request) contains only thirty lines of code. One can use Stassen's algorithm to compute multiplication of matrices and the binary algorithm to calculate the power of matrices. Thus the program would be of complexity $a^{2,7} \cdot \log n$, with n the length of the sequence corresponding to the power of the matrix, and a the calculated local score corresponding to the size of the matrix ($a + 1$ more precisely).

The speed and the precision are obviously better for short sequences but are surprisingly good even for long ones. One may control the computational error using the fact that the sum of each row of Π^n must be equal to one. For very small (for example $< 10^{-12}$) P-values, the approximation may be preferred to the exact result.

Table 1 compares numerical results on real sequences using the method of Theorem 1 with the results of Karlin *et al.* in [5]. The indices in the table are the same as the ones used for the examples in [5]. The comparison is done using the relative error.

Table 1. Comparison of exact and asymptotic values of $P(H_n \geq a)$ on Karlin and Altschul examples: a) research for high-scoring mixed charge segments (of basic and acidic residues); b) research for high-scoring acidic charge segment; c) high-scoring basic charge segment; d) Strong hydrophobic segments; e) Cysteine cluster.

Index	X	n	a	KA P-value	Exact P-value	Relative Error
a) (i)	2 and -1	643	21	$< 8.10^{-3}$	$6,99.10^{-3}$	14%
a) (ii)	2 and -1	331	29	$< 2.10^{-4}$	$1,04.10^{-4}$	92%
b)	2, -1 and -2	575	11	$3,7.10^{-3}$	$2,75.10^{-3}$	34%
c) (i)	2, -1 and -2	1320	10	$3,4.10^{-2}$	$3,56.10^{-2}$	-5%
c) (ii)	2, -1 and -2	575	12	4.10^{-3}	$3,97.10^{-3}$	0,7%
c) (iii)	2, -1 and -2	614	37	$< 2.10^{-4}$	$9,22.10^{-5}$	117%
d) (i)	1, -1 and -2	552	17	$1,8.10^{-5}$	$1,73.10^{-5}$	4%
d) (ii)	1, -1 and -2	325	15	8.10^{-2}	$7,47.10^{-2}$	7%
d) (iii)	1, -1 and -2	1480	21	10^{-3}	$9,31.10^{-4}$	7%
e)	5 and -1	575	12	$9,1.10^{-1}$	$9,50.10^{-1}$	-4%

Table 2. Comparison between values of Daudin and Mercier (DM), asymptotic values of Karlin *et al.* (KA), and empirical values (Emp.) of $P(H_n \geq a)$ on the examples of Karlin and Altschul [5]. Empirical distribution has been calculated on 10^7 or 10^6 simulated sequences.

Index	P-value			Relative Error		
	KA	DM	Emp.	DM/KA	KA/Emp.	DM/Emp.
a) (i)	$< 8.10^{-3}$	$6,99.10^{-3}$	$7,12.10^{-3}$	14%	12,3%	-2%
a) (ii)	$< 2.10^{-4}$	$1,04.10^{-4}$	$1,04.10^{-4}$	92%	92%	0%
b)	$3,7.10^{-3}$	$2,75.10^{-3}$	$2,71.10^{-3}$	34%	36%	1,5%
c) (i)	$3,4.10^{-2}$	$3,56.10^{-2}$	$3,56.10^{-2}$	-5%	-5%	0%
c) (ii)	4.10^{-3}	$3,97.10^{-3}$	$3,94.10^{-3}$	0,7%	1%	0,7%
c) (iii)	$< 2.10^{-4}$	$9,22.10^{-5}$	$9,1.10^{-5}$	117%	119%	1,3%
d) (i)	$1,8.10^{-5}$	$1,73.10^{-5}$	$1,8.10^{-5}$	4%	0%	4%
d) (ii)	8.10^{-2}	$7,47.10^{-2}$	$7,50.10^{-2}$	7%	7%	0,4%
d) (iii)	10^{-3}	$9,307.10^{-4}$	$9,30.10^{-4}$	7%	7%	0,07%
e)	$9,1.10^{-1}$	$9,50.10^{-1}$	$9,50.10^{-1}$	-4%	-4%	-0,04%

We can see that there are two important relative errors for a)(ii) and c)(iii). These values have been confirmed by simulation (see Table 2). We also confirm the fact that the Karlin *et al.* formula brings a negative error for example c)(i) and e), whereas the authors proposed a superior threshold in order to be conservative.

For very small P-values (for example $< 10^{-12}$) it would be more appropriate to consider other indicators such as a Z-score for example.

$$Z = (a - E[H_n])/\sigma[H_n] \ . \tag{13}$$

Z-score is already used but the difficulty consists in establishing the mean and the variance of the local score. Solutions are proposed using bootstrap methods. This problem does not exist with the method we propose. Using Formula (12), the exact mean and variance of H_n can be easily computed. Therefore Formula (13) is accessible and may be easier to interpret than a P-value. A Z-score measures the number of $\sigma[H_n]$ where the calculated value a differs from the mean: two Z-scores of 8 and 10 are probably more easily compared than their /tmptwo corresponding P-values of respectively 10^{-22} and 10^{-24}.

3.2   Alignment of Two Sequences with Shift

The case of the local score for alignment of two sequences is more intricate than the case of one sequence: one may easily adapt the preceding result to the case of alignment between two sequences with shift but without gap. This has been written in [9]. Here are the main steps of the generalization.

For the ungapped alignment problem, considering shifts (see [10]), we can use the exact probabilities in the following way. Let us consider two sequences of size m and n, respectively, with $m \geq n$. We can align the first sequence with $m + n - 1$ sequences obtained by slipping the second sequence (see Fig. 3).

The local score is thus defined as

$$H_{n,m} = \max_{\substack{0 \leq \ell \leq \min(n,m)-1, \\ 1 \leq i \leq n-\ell \\ 1 \leq j \leq m-\ell}} \sum_{k=0}^{\ell} \sigma(A_{i+k}, B_{j+k})$$

where $A_1...A_n$ and $B_1...B_m$ correspond to the sequences and σ is the scoring function reflecting similarities between the components (see [3] for amino acids). Like most authors we do not take into account the dependence between them.

Therefore the probability of exceeding a is approximatively equal to

$$P(H_{n,m} \geq a) = 1 - \prod_{i=1,...,n-1} P(H_i < a)^2 P(H_n < a)^{m-n+1}$$

Note that we are able to take into account the size of each pair of sequences, which is not possible when an asymptotic method is used.

Numerical comparisons on real examples have not been done yet.

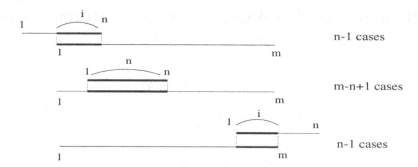

Fig. 3. Alignment of two sequences with shift. There is a total of $m + n - 1$ different positions between the two sequences that can be regrouped in three main cases.

4   Perspective and Conclusion

Further work comes naturally. The results on new asymptotic and exact distribution of the local score are established with independent and identically distributed random variables. The next step would be to generalize these results with the Markovian model, which takes more information from the sequences. Some work must be made to adapt them to the case of alignment with gaps.

References

1. Altschul, S., Gish, W., Miller, W. Myers, E., Lipman, D.: Basic Local Alignment Search Tool. J. Mol. Biol. **215** (1990), 403–410
2. Daudin, J.-J., Mercier, S.: Distribution exacte d'une suite de variables indépendantes et identiquement distribuées. C. R. Acad. Sci. Paris tome 329 série I (1999) 815–820
3. Dayhoff, M., Schwartz, R., Orcutt, B.: A model of evolutionary change in protein. Atlas of Protein Sequences and Structure, **5** (1978), 345–352
4. Dembo, A., Karlin, S., Zeitouni, O.: Limit distribution of maximal non-aligned two-sequences segmental score. Ann. Prob. **22** (1994) 2022–2039
5. Karlin, S., Altschul, S.F.: Methods for assessing the statistical significance of molecular sequence features by using general scoring schemes. Proc. Natl. Acad. Sci. USA **87** (1990) 2264–2268
6. Karlin, S., Dembo, A.: Limit distributions of maximal segmental score among Markov-dependent partial sums. Adv. Appl. Prob. **24** (1992) 113–140
7. Karlin, S., Dembo, A., Kawabata, T.: Statistical composition of high-scoring segments from molecular sequences. Ann. Statist. **18** (1990) 571–581
8. Mercier, S.: Statistiques des scores pour l'analyse et la comparaison de séquences biologiques. Thèse, Université de Rouen (1999)
9. Mercier, S., Daudin, J.-J.: Exact distribution for the local score of one i.i.d. random sequence. To appear in J. Comp. Biol.
10. Mott, R., Tribes, R.: Approximate Statistics of Gapped Alignments. J. Comp. Biol. **6** 1 (1999) 91–112

Phylogenetic Reconstruction Algorithms Based on Weighted 4-Trees

Vincent Ranwez and Olivier Gascuel

LIRMM, UMR 9928 UNIVERSITE MONTPELLIER II/CNRS
161, Rue Ada
34392 Montpellier cedex 5 France
{ranwez,gascuel}@lirmm.fr
http://www.lirmm.fr/~w3ifa/MAAS/

Abstract. Quartet methods first compute 4-taxon trees (or 4-trees) then use a combinatorial algorithm to infer a phylogeny that closely respects the inferred 4-trees. This article focuses on the special case involving weighted 4-trees. The sum of the weights of the 4-trees induced by the inferred phylogeny is a natural measurement of the fit between this phylogeny and the 4-tree set. In order to measure the fit of each edge of the inferred tree, we propose a new criterion that takes the weights of the 4-trees along with the way they are grouped into account. However, finding the tree that optimizes the natural criterion is NP-hard [10], and optimizing our new criterion is likely not easier. We then describe two greedy heuristic algorithms that are experimentally efficient in optimizing these criteria and have an $O(n^4)$ time complexity (where n is the number of studied taxa). We use computer simulations to show that these two algorithms have better topological accuracy than *QUARTET PUZZLING* [12], which is one of the few quartet methods able to take 4-trees weighting into account, and seems to be widely used.

1 Introduction

The maximum-likelihood method [4] is widely used to infer molecular phylogenies. It has sound statistical foundations and performs well in computer simulations. Unfortunately, the computing time required quickly becomes unacceptable as the number n of taxa increases. When n is greater than about a dozen taxa, it is much faster to use maximum-likelihood to analyze each subset of four taxa than to use it for directly inferring the evolutionary history of all of these taxa. Based on these likelihood values, quartet methods give a weight to every 4-taxon phylogeny (4-tree), and rely on a combinatorial algorithm that searches for the n-tree which best fits this set of weighted 4-trees. A natural criterion to measure the fit between a phylogeny and the set of weighted 4-trees is to sum up the weights of the 4-trees induced by this phylogeny. We propose a new criterion in this article, which takes into account both the weights of the 4-trees and the way they are grouped.

O. Gascuel, M.-F. Sagot (Eds.): JOBIM 2000, LNCS 2066, pp. 84–98, 2001.

In this context, quartet methods are used to simplify the original maximum-likelihood approach in order to benefit from its strength within reasonable computing time. However, combining 4-trees is computationally difficult, since deciding whether a given set of 4-trees can be combined into a phylogeny is an NP-complete problem [10], and the associated optimization problem (*i.e.* finding the phylogeny that satisfies the maximum number of 4-trees) is NP-hard. Moreover, weighting the 4-trees seems to improve the performance of quartet methods [11], but it makes optimization harder. Concerning the new criterion defined and studied in this article, it is very likely that the optimization task is not easier (even though the formal proof would not be given here). For these two criteria (and others), we thus have to rely on heuristic algorithms which provide acceptable solutions in reasonable computing time.

QUARTET PUZZLING [12,11] relies on such heuristic algorithm, and currently seems to be the most widely used quartet method. Starting from a phylogeny of three taxa, *QUARTET PUZZLING* (*QP*) progressively constructs a complete phylogeny by successively adding the remaining taxa. For this purpose, it optimizes a local criterion that indicates the best edge to insert the new taxon. Since the resulting phylogeny depends on the taxa addition order, *QP* constructs 1000 trees using a different random addition order for each tree. Finally, *QP* proposes the majority rule consensus [8] of these 1000 trees. As previously described [12], *QP* has $O(n^5)$ time complexity. Although it is possible to implement it in $O(n^4)$, the need to construct 1000 trees is highly penalizing in terms of computing time and memory space.

In this article, we introduce two heuristic algorithms that construct a single n-tree, which optimizes a global criterion. We first describe the most natural criterion to measure the fit between a phylogeny and a set of weighted 4-trees, and then introduce our new criterion (Sect. 2). Based on these two criteria, we propose (Sect. 3) two new phylogenetic reconstruction algorithms, and describe (Sect. 4) an $O(n^4)$ implementation for one of them, which can be extended to the other. Finally, we use computer simulations to compare the topological accuracy of these two algorithms with that of *QP* (Sect. 5).

2   A Criterion Based on 4-Tree Partitioning

First we define the terms employed in this article and provide the corresponding notation. Then we describe the natural criterion to measure the fit between a phylogeny and a set of weighted 4-trees. Finally, we describe a new criterion for measuring this fit and explain why we introduce this new criterion.

2.1   Definitions and Notations

A tree is a set of nodes connected by edges (or branches) so that two distinct nodes are connected by a single path. The degree of a node is the number of edges attached to it. The nodes with degree 1 (the leaves) are labelled by taxa; all other nodes are called internal nodes. If all the internal nodes are of degree

3, then the tree is said to be fully resolved; if some have a higher degree, we say that the tree is partially resolved. An edge connecting two internal nodes is an internal edge.

Removing an edge in a phylogeny separates taxa into two disjointed subsets with union equal to the complete set of taxa. The edge is said to define the split (or bipartition) formed from the two subsets.

Let $xy|zt$ denote the 4-tree that separates taxa x and y from taxa z and t. A quartet is a set of 4 taxa; for each quartet $\{x, y, z, t\}$ there are three possible 4-trees: $xy|zt$, $xz|yt$ and $xt|yz$. Let the pair (q, w) denote the weighted 4-tree (w4-tree) which associates weight w with the 4-tree q. When an n-tree T contains a split separating taxa x and y from the taxa z and t, T is said to induce the 4-tree $xy|zt$ and the 4-tree $xy|zt$ is said to be consistent with T.

Let Q denote the set of weighted 4-trees used as starting point by quartet methods; we assume that Q contains three w4-trees for each quartet. The set of 4-trees induced by a tree T is denoted Q_T. After assigning weights to the 4-trees, quartet methods search for the tree T which fits them best. A natural measurement of this fit is the sum of the weights of the 4-trees induced by T [2]. Thus, we search for the tree T which maximizes the W criterion defined as:

$$W(T) = \sum_{q \in Q_T \text{ and } (q,w) \in Q} w \ . \tag{1}$$

However, finding T thus defined is NP-hard [10] and can require an exponentially long computing time. Therefore, heuristic algorithms such as QP have to be used to find a near-optimal tree within a reasonable amount of computing time.

2.2   A New Criterion

A fully resolved phylogeny T on a set of taxa E induces a single 4-tree for each quartet of E. The internal edge of each of these 4-trees corresponds to a path in T. Thus, gathering 4-trees whose internal edge corresponds to the same path in T provides a partition of the 4-trees. This allows us to define a weight for each path corresponding to the average weight of its associated 4-trees. The sum of weights of paths from T defines a new criterion denoted as P.

In the example given in Fig. 1, the phylogeny T on the five taxa $\{1, 2, 3, 4, 5\}$ has a W criterion value of 4.5 and a P criterion value of 2.75; $i.e.$ the average weight of a w4-tree is 0.9 whereas the average weight of a path is about 0.92. Indeed, W is the weight sum of the 5 w4-trees:

$$W(T) = 1/2 + 1 + 1 + 1 + 1$$

whereas P is the weight sum of the 3 paths (n_1, n_2), $(n_2 n_3)$ and $(n_1 n_3)$:

$$P(T) = \frac{1/2 + 1}{2} + \frac{1 + 1}{2} + \frac{1}{1}.$$

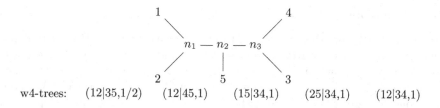

w4-trees: $(12|35,1/2)$ $(12|45,1)$ $(15|34,1)$ $(25|34,1)$ $(12|34,1)$

Fig. 1. Difference between W and P: a simple example.

2.3   Why Introduce a New Criterion?

Both W and P are reasonable criteria. Indeed, in the case where all 4-trees induced by the correct phylogeny have a weight equal to 1 and all others a weight equal to 0, both criteria have the correct phylogeny as single optimum. Considering W, the correct phylogeny has a weight equal to the number of quartets. Whereas all other phylogenies induce at least one incorrect 4-tree and thus has a lower W value. Considering P, every fully resolved phylogeny on n taxa has the same number $(n-2)$ of internal nodes and thus the same number (C_{n-2}^2) of paths. Therefore, the correct phylogeny has a weight equal to the number of paths, whereas any other phylogeny induces at least one incorrect 4-tree, and thus at least one of its paths has a weight of less than 1.

An internal edge of a fully resolved phylogeny induces 2 subtrees on one side (denoted as A_1 and A_2) and 2 others on the other side (denoted as B_1 and B_2). The 4-trees $a_1a_2|b_1b_2$ such as $a_1 \in A_1$, $a_2 \in A_2$, $b_1 \in B_1$ and $b_2 \in B_2$ are said to be *specific* to this edge while the 4-trees $a_1a_2|b_1b_2$ such as $a_1 \in A_1\bigcup A_2$, $a_2 \in A_1\bigcup A_2$, $b_1 \in B_1\bigcup B_2$ and $b_2 \in B_1\bigcup B_2$ are said to be *induced* by this edge. For a fully resolved phylogeny, some edges have much more specific 4-trees than others. Considering only the sum of the weights of the 4-trees focuses on the resolution of edges with many specific 4-trees but to the detriment of others. Defining the phylogeny weight as the sum of weights of its edges could avoid this bias. Unfortunately, weighting edges through their specific 4-trees ignores some 4-trees, whereas by weighting them through the 4-trees they induce, some 4-trees are used many more times than some others. In practice, both weighting methods tend to favor some topologies. This led us to define a weight for the paths rather than for the edges. The weight of a w4-tree is thus taken into account exactly once.

3   Weight Optimization and Path Optimization

WEIGHT OPTIMIZATION (*WO*) and *PATH OPTIMIZATION* (*PO*) are both greedy algorithms that optimize the W and the P criterion, respectively. Both algorithms have an $O(n^4)$ time complexity, and construct a single phylogeny.

3.1   Principle

WO and *PO* follow the same algorithmic scheme. They use the w4-tree with maximal weight as a starting point and then add one taxon at a time until a complete *n*-tree containing all taxa is obtained. Each time we add a new taxon, we can add it onto any of the edges of the partially constructed tree. We choose the edge that gives the tree with the largest criterion. Unlike quartet puzzling, the order that we use for adding taxa is not determined randomly. Instead, at each step we add the taxon providing the "safest" addition, and thus use the 4-tree weights to dynamically define the taxa addition order.

We define the safety $s(i)$ of the addition of i as follows. Let M denote the edge with the highest bonus and m the edge giving the second highest bonus. Let $\delta W(M, i)$ and $\delta W(m, i)$ denote the increase in $W(T)$ resulting from adding i onto edge M or m. We have:

$$s(i) = \frac{\delta W(M, i) - \delta W(m, i)}{\delta W(M, i) + \delta W(m, i)} , \tag{2}$$

and the definition of s is identical for P. This definition of safety differs from that of [13] which uses $s(i) = \delta W(M, i) - \delta W(m, i)$. Unpublished tests show that in our case it is preferable to normalize the safety between 0 and 1 as done in the above definition.

The $O(n^4)$ implementations of these algorithms are obtained by gathering w4-trees that modify bonuses of the same edges and by reusing calculations already carried out in previous steps. *WO* is the easiest to implement, and we will thus describe it in detail. Implementation of *PO* is based on the same ideas.

3.2   Edge Bonus Computation

Let T denote the current partially constructed tree. In order to evaluate the safety of the addition of a new taxon i on T, *WO* considers all w4-trees of type $(ix|yz, w)$ (with x, y and z in T), which are the "relevant" w4-trees when i is added. For each of these w4-trees, *WO* adds a bonus w to each edge of the current tree, such that the addition of i constructs a tree inducing $ix|yz$. Once all relevant w4-trees have been taken into account, the bonus of any edge e is equal to the increase $\delta W(e, i)$ of the W criterion induced by the addition of i onto e. Considering the w4-tree $(ix|yz, w)$, the edges receiving a bonus are determined as follows. There is a single internal node belonging to the three paths (x, y), (x, z), and (y, z), called the median of x, y, z and denoted as $m(x, y, z)$ [1]. This internal node is attached to three edges and thus defines three disjointed sub-trees denoted as T_x, T_y and T_z, such that $x \in T_x$, $y \in T_y$ and $z \in T_z$ (Fig. 2). Every edge of T (as well as every leaf of T) belongs to a single one of these subtrees. If taxon i is added on an edge of T_x, then all edges of the path (i, x) are within T_x while those of the path (y, z) are within $T_y \bigcup T_z$. This ensures that paths (i, x) and (y, z) do not intersect and consequently that the constructed tree induces $ix|yz$. Thus, considering $(ix|yz, w)$, *WO* adds bonus w to the edges of T_x.

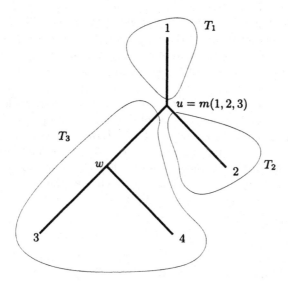

Fig. 2. Median node. The median node of 1, 2 and 3 denoted $m(1,2,3)$ splits the tree into three subtrees T_1, T_2 and T_3. Every edge of T belongs to a single one of these subtrees.

3.3   Basic $O(n^4)$ Implementation of WO

Let T denote the current partially constructed tree. An internal node of T splits it into three sub-trees denoted as T_x, T_y and T_z (Fig. 2). Gathering the relevant quartets $\{i, x, y, z\}$ such that $x \in T_x$, $y \in T_y$, $z \in T_z$ and i is a taxon not yet added, allows an efficient computation of edge bonuses relative to the addition of i. Indeed, instead of parsing T_x for each w4-tree $(ix|yz, w)$ in order to add a bonus w on all edges of T_x, it is more efficient to compute the sum S of these bonuses and to add it on the edges of T_x by a single tree search of T_x. When a new taxon z_{new} is added to one of these sub-trees, say in T_z, S can be updated just by adding to it the weights of the new w4-trees of type $(ix|yz_{new}, w)$. The WO implementation (Algorithm 1) exploits these observations. It associates a value denoted as $S_i(T_x)$ to each sub-tree T_x and for each not yet added taxon i. These $O(n^2)$ values are updated after each new addition.

The algorithm described below initializes the result tree T with the 4-tree of maximum weight. This step requires consultation of every w4-tree, and thus has an $O(n^4)$ time complexity. Then it initializes the set R of remaining not yet added taxa and stores the last one added. Thereafter, while R is not empty, each remaining taxon is considered and the associated bonuses are computed; the taxon with the highest safety is added to T on the edge having the highest bonus for this taxon. This taxon is removed from R and becomes the new latest added taxon. For a particular taxon i, the edge bonuses are computed as follows. First, they are set at 0; then, each internal node is considered and (loop 1) the values to propagate for taxon i in the three corresponding sub-trees are updated

to take the latest added taxon into account ; then (loop 2) these three values
are added to the bonuses of the edges belonging to the corresponding sub-trees.
Note that the addition of a new taxon generates four new subtrees whose initial
null S values are updated as well as others during the next passage through
loop 1.

By removing loop 1, we clearly obtain an $O(n^4)$ algorithm (four nested loops
of n steps). By removing loop 2, we obtain an algorithm which consults each
4-tree weight only once, and hence is also an $O(n^4)$ algorithm. This, in addition
to the fact that loop 1 and loop 2 are executed sequentially, ensures that the
whole complexity of the WO algorithm is $O(n^4)$. In our experiments, we use a
refined implementation of this algorithm which uses the tree search introduced
in [3] to propagate bonuses along subtrees.

---

input : A set Q of w4-trees corresponding to the n taxa
output : A phylogeny of this n taxa

find the 4-tree $ab|cd$ of maximum weight;
initialize T with the tree (a, b, c);
foreach *taxon i of $\{1, 2, ..., n\} - \{a, b, c\}$* **do**
 add the weight of $(ia|bc)$ to $S_i(T_a)$;
 add the weight of $(ib|ac)$ to $S_i(T_b)$;
 add the weight of $(ic|ab)$ to $S_i(T_c)$;
add d to the edge of T connected to taxon c;
$R \leftarrow \{1, 2, ..., n\} - \{a, b, c, d\}$;
$last \leftarrow d$;
while $R \neq \emptyset$ **do**
 foreach *taxon i of R* **do**
  reinitialize edge weights to 0 ;
  foreach *internal node of T splitting T in $T_{last} \ni last$, T_x, T_y* **do**
   (1) **foreach** $(x, y) \in T_x \times T_y$ **do**
    add the weight of $(ix|ylast)$ to $S_i(T_x)$;
    add the weight of $(iy|xlast)$ to $S_i(T_y)$;
    add the weight of $(ilast|xy)$ to $S_i(T_{last})$;
   (2) **foreach** *sub-tree $T_z \in \{T_x, T_y, T_{last}\}$* **do**
    add $S_i(T_z)$ to the bonus of each edge of T_z ;
  memorize the best edge for taxon i and its safety;
 $last \leftarrow$ the taxon providing the safest addition;
 add $last$ on its best edge and remove it from R;
return(T);

Algorithm 1: The *WEIGHT OPTIMIZATION* algorithm.

3.4   Memory Space Required

In Algorithm 1, the size of the input data is $O(n^4)$ and, as for most quartet algorithms, the memory space required is thus $O(n^4)$. However, it is possible to modify quartet methods in order to input the DNA sequences and to compute the weight of a 4-tree each time it is needed. This decreases the memory space required but increases the computing time as soon as weights are needed more than once. QP needs to consult each weight at least 1000 times (one per reconstructed tree). For such methods, recomputing the weights instead of storing them is not realistic. However, this approach is better adapted in our case, since in Algorithm 1 the weight of a 4-tree is consulted only twice, first to determine the starting point and second when it becomes a relevant 4-tree. Moreover, there is no real need to start with the best 4-tree, we just need a good one. A solution is to randomly select the three taxa a, b, c used to initialize the tree T, and the next taxon added will be the one with the highest safety. This ensures that the starting 4-tree is a good one, since it is the best among the $3 * (n-3)$ 4-trees that resolve a quartet $\{a, b, c, x\}$, with x differing from a, b and c. In practice, this modification has no significant influence on the topological accuracy of the algorithm and saves memory space and computing time. Denoting l as the length of DNA sequences, the memory space required is then $O(nl)$ for storing the initial DNA data, $O(n^2)$ to store the value to propagate in each subtree for each taxon, $O(n^2)$ to store the leaves of each subtree. The remaining memory required is $O(n)$, so that the whole space complexity is $O(nl + n^2)$. This relatively low space complexity allows us to deal with larger data sets than quartet methods that need to store the $O(n^4)$ weights.

3.5   Optimizing the Propagations of Bonuses

The previous algorithm can be improved by propagating values in all subtrees at a time rather than for each internal node. Once all $S_i(T)$ have been updated, they may be propagated in the corresponding subtree through only two tree searches. This particular use of tree search described in [3] is detailed below, where i is omitted for the sake of simplicity.

A leaf r of T is selected to root the current tree, this choice induces an orientation of T. An internal node u of T splits T into three subtrees, one of which contains r and is called the upper tree and denoted U_u, while the two others are called left subtree L_u and right subtree R_u. Note that when u is a leaf, it is associated to only one rooted (sub)tree which is the current tree T with u as root; this (sub)tree is denoted R_u if u is the root of T and U_u otherwise. Note that when u is a leaf $S_i(U_u)$ (or $S_i(R_u)$) is consistently set at 0.

Basically, a subtree inherits the initial scores from all subtrees containing it. For example, in Fig. 3, both L_w and R_w inherit from L_u, and therefore their final scores are respectively 2 and 1.8. For each internal node, three recursive relations link each of the three subtrees with the two subtrees that immediately contain this subtree. For example, in Fig. 3 we have the relations:

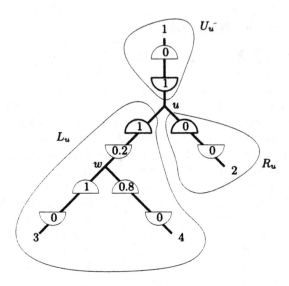

Fig. 3. The already existing four-species tree 12|34 and the initial scores of the rooted subtrees when the w4-trees involving taxon 5 are: (51|23, 1), (53|14, 1), (54|12, 1), (54|23, 0.8) and (52|34, 0.2). For example, the initial scores of U_u, L_u and R_u are respectively 1, 1 and 0; the first 1 comes from (51|23, 1) and the second from (54|12, 1).

$$S(U_w) \leftarrow S(U_3) + S(U_4),$$
$$S(L_w) \leftarrow S(L_u) + S(U_4) \text{ and}$$
$$S(R_w) \leftarrow S(L_u) + S(U_3).$$

The proposed procedure exploits these "unoriented" recurrence relations. It starts by the larger subtrees, *i.e.* the upper trees of the leaves, and carefully navigates through the tree so that a subtree is processed only when the subtrees that contain it have already been processed. This is achieved using a double recursive search [3] to deal with such unoriented recurrence relations. The first recursive search is depth-first and post-ordered. It computes the score of every upper subtree using the scores already obtained for the upper subtrees of its "children". This post-ordered search (Algorithm 2) is run on the only child of T's root (see also Fig. 4).

PostOrder(u)
if u *is a leaf* **then** return 0;
else
    | let l be the left-children of u and r its right-children:
    | $S(U_u) \leftarrow S(U_u) + \text{PostOrder}(l) + \text{PostOrder}(r)$;
return $S(U_u)$;

Algorithm 2: *The post-ordered search.*

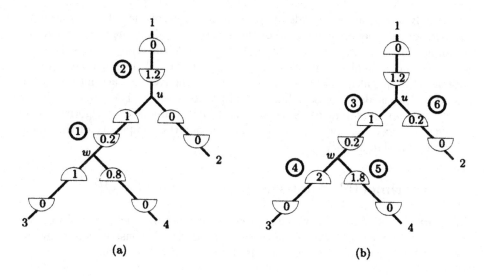

Fig. 4. Propagation of scores in the tree in Fig. 3, by the PostOrder procedure (a) and the PreOrder procedure (b). The processing order of the subtrees is indicated by the circled numbers. The selected edge is $(3, w)$, the weight of which is $2 + 0 = 2$.

The second recursive search is depth-first and pre-ordered. It computes scores of the lower (left or right) subtrees, using scores of the upper subtrees (computed during the previous search) and scores of the lower subtrees (computed during the previous recursive calls). This procedure (Algorithm 3) takes two arguments: the only child of T's root, and the value 0 that corresponds to the weight of the lower subtree attached to T's root (see also Fig. 4).

PreOrder(u,s)
if u *is an internal node* **then**
  | let l be the left-children of u and r its right-children:
  | $S(L_u) \leftarrow S(L_u) + S(U_r) + s$;
  | PreOrder $(l,S(L_u))$;
  | $S(R_u) \leftarrow S(R_u) + S(U_l) + s$;
  | PreOrder $(r,S(R_u))$;
else
  | stop;

Algorithm 3: *The pre-ordered search.*

Finally, the score of any edge (n_1, n_2) is the score of the subtree containing n_1 but not n_2, added to the score of the subtree containing n_2 but not n_1. For example (Fig. 4), the score of (u, w) is $1 + 0.2$, while the score of $(1, u)$ is $0 + 1$.

Note that each of the two recursive searches has $O(n)$ time complexity, thus instead of propagating the S_i value by searching the tree for each internal node $(O(n^2))$, this is done by searching it only twice $(O(n))$. The same holds for every remaining taxon i, so this replaces an $O(n^4)$ component of Algorithm 1 by an $O(n^3)$ procedure. Although it does not change the complexity of the whole algorithm, it significantly speeds it up.

4   Experimental Comparison of WO, PO and QP

We start by describing how we generated our test sets and the protocol we used to compare the performance of the various phylogenetic reconstruction programs and criteria. Then we comment on the results obtained from the different methods.

4.1   Protocol

Our experimental tests follow a protocol used within a similar framework in [7] and later in [6]. Three of the model trees used for these tests respect the molecular clock, and three do not. We studied three cases for each of these model trees (Fig. 5):

- a low evolution rate, for which the Maximum Pairwise Divergence (MD) is about 0.1 substitution per site ($a = 0.0055$ and $b = 0.004$),
- a medium evolution rate, $MD \simeq 0.3$ ($a = 0.0165$ et $b = 0.012$),
- a fast evolution rate, $MD \simeq 1$ ($a = 0.055$ et $b = 0.04$).

For each tree T, we used $SEQGEN$ [9] to generate 1000 data files containing sequences of length 300. These sequences were obtained by simulating, along T, an evolving process according to the Kimura two parameter model with a transition/transversion rate equal to 2. We thus tested the different methods on 18000 test sets corresponding to 3 evolution rates and 6 model trees.

The phylogenetic reconstruction methods were assessed on the basis of their ability to infer the correct tree used to generate the sequences. This evaluation was done by counting how many times the tree proposed by the method is the correct tree.

4.2   Methods Tested

The methods tested are WO, PO and a close variant of the most recent version (4.2) of Quartet Puzzling QP. By default, Quartet Puzzling constructs 1000 trees (as described above) and returns their majority rule consensus tree [8]. Instead of this majority rule consensus tree, we use the greedy consensus tree

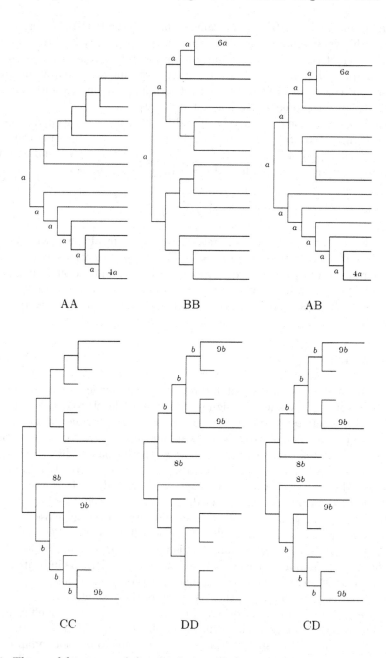

Fig. 5. The model trees used for simulation. Each interior branch is one unit long (a for constant and b for variable-rate trees) and the length of external branches are given in multiples of a or b. Low divergence refers to $a = 0.0055$ and $b = 0.004$ which corresponds to a maximum pairwise divergence (MD) of about 0.1 substitution per site. Medium divergence refers to $a = 0.0165$ and $b = 0.012$ ($MD \simeq .3$) and high divergence refers to $a = 0.055$ and $b = 0.04$ ($MD \simeq 1$).

obtained from the *CONSENSE* program available in the PHYLIP package [5]. In our tests, this version of Quartet Puzzling provides fully resolved trees and thus infers the correct tree more often than the majority rule consensus, whose trees are usually partially unresolved. Moreover, this variant was used by the authors of Quartet Puzzling in their tests [12,11] (personal communication). The three methods balance the quartets as suggested in [11].

Given a quartet, the likelihood l_1, l_2 and l_3 of the three associated 4-trees are used to estimate their respective probabilities p_1, p_2 and p_3 of being the correct 4-tree, and these probabilities are then used to weight the corresponding 4-trees. These probabilities can be evaluated using Bayes' theorem as:

$$p_i = \frac{l_i}{l_1 + l_2 + l_3} \, . \tag{3}$$

The weights w_i used by *QP* are discrete approximations of the p_i with values in $\{0, 1/2, 1/3, 1\}$ such that $w_1 + w_2 + w_3 = 1$. These approximations do not degrade the performance of *QP* but involve more efficient computation in comparison to the original real valued weight [11]. These approximations are not used for *WO* and *PO* because unpublished simulations show that they degrade their performance.

4.3   Results

Table 1 indicates the percentage of inferred trees which exactly correspond to the correct tree topology. We use the sign test (at the 5% level) to decide whether or not a difference between two results is significant. We underline the worst results and put it in bold when the difference is significant relative to the two others.

Table 1 indicates that *QP* has often worst performance than *WO* and *PO* (14 cases over the 18). *QP* actually performs well with the tree BB. We assume that this is due to the use of a consensus method. Indeed, it is easy to verify that if 1000 trees are randomly labeled, even though their topology is that of AA, the greedy consensus tree has an equilibrated topology close to that of BB. Conversely, it is not easy to determine which of *WO* or *PO* has the best performance. Table 1 indicates that *WO* is more efficient than *PO* for the tree AA, whereas *PO* is more efficient than *WO* for the tree BB. As indicated above, if the ratio between the maximal number and the minimal number of 4-trees specific to an edge is too large, then *W* favours the resolution of certain parts of the tree to the detriment of others. In such cases, *PO* is more efficient than *WO*. This situation corresponds to the tree BB with ratio 81/9. For other trees, the results depend on the substitution rate. *PO* seems to be more adapted than *WO* for a maximal pairwise divergence of 0.3, which is maybe the tested value closest to those generally observed when phylogenetic reconstruction methods are used.

It would seem that the results obtained by *WO* and *PO* could be improved by better optimization of their respective criteria. Considering *WO*, when the correct Tree T is not inferred by *WO*, two cases arise. Either T is the single

Table 1. Percentage of Well Reconstructed Trees by *QP*, *WO* and *PO*

evolution rate	slow			medium			fast		
Tree \ Algorithm	*QP*	*WO*	*PO*	*QP*	*WO*	*PO*	*QP*	*WO*	*PO*
AA	**5**	11	11	**15**	31	25	**6**	25	14
BB	20	**9**	17	43	**26**	40	36	**22**	31
AB	**7**	8	10	**24**	28	31	**15**	24	22
CC	7	**7**	8	**35**	40	42	**39**	60	52
DD	**6**	7	8	**31**	40	40	**44**	59	52
CD	**7**	8	8	**36**	42	43	**42**	61	52

optimum according to W, and in this case the problem is the result of WO incorrectly optimizing the W criterion. Or T is non-optimal (or not the single optimum), and in this case the problem comes from the W criterion which is not discriminating enough. The same holds for PO and P.

In order to know which of these two problems limits the performance of WO (resp. PO) , we isolated the second kind of error. Instead of defining a dynamic addition taxon order, we used 100 random addition orders and constructed the 100 corresponding trees. Then among these 100 trees, we kept the one which had the highest criterion value. This algorithm, denoted as WO_{100} (resp. PO_{100}), does not guarantee that the tree returned is optimal. However, in our tests, it always inferred a tree whose criterion is at least as good as that of the correct tree T. Thus, the only errors made by WO_{100} (resp. PO_{100}) correspond to cases where the W criterion (resp. the P criterion) is not sufficiently discriminating. Then the performance of WO_{100} and PO_{100} also allows us to compare the pertinence of the two criteria.

The obtained results (Table 2) are similar to those observed for WO or PO. The results of WO_{100} (resp. PO_{100}) are hardly better than those of WO (resp. PO). The greatest improvement is less than 6% for PO (tree BB, high evolution rate) and 3% for WO (tree BB high evolution rate) and in both case the average improvement is about 1.5%. It seems that, due to its complexity, the P criterion is a little bit more difficult to optimize, but it also appears that only a tiny part of the errors are due to insufficient optimization.

5   Conclusion

We introduced a new criterion to measure the fit between a set of weighted 4-trees and a phylogenetic tree on the same set of taxa. We have experimentally shown that this criteria, is at least as relevant as the sum of the weight of the 4-trees. Based on these criteria we have introduced two new phylogenetic reconstruction algorithms with $O(n^4)$ time complexity. They both construct a single phylogeny and thus are faster than Quartet Puzzling (which need to construct 1000); moreover, they have a better topology accuracy. However, quartet

Table 2. Percentage of times the quasi-optimum of the criterion is the correct tree

evolution rate	slow		medium		fast	
Tree \ Algorithm	WO_{100}	PO_{100}	WO_{100}	PO_{100}	WO_{100}	PO_{100}
AA	11	__11__	31	__25__	26	__14__
BB	__10__	18	__27__	43	__25__	37
AB	__9__	10	__29__	32	27	__23__
CC	__7__	9	__41__	43	60	__53__
DD	__7__	9	__40__	43	61	__56__
CD	__8__	9	__42__	46	61	__54__

methods are likely penalized due to their lack of global vision, and the consequences of the local approach adopted by quartet methods have not yet been studied.

References

[1] J. P. Barth lemy and A. Gu noche. *Trees and Proximity Representation.* John Wiley & sons, 1990.

[2] V. Berry. *M thodes et algorithmes pour reconstruire les arbres de l' volut ion.* PhD thesis, Univ. Montpellier II, d cembre 1997.

[3] V. Berry and O. Gascuel. Inferring evolutionary trees with strong combinatorial evidence. *Theoretical Computer Science*, 240:271–298, 2000.

[4] J. Felsenstein. Evolutionary trees from DNA sequences: a maximum likelihood approach. *J. Mol. Evol.*, 17:368–376, 1981.

[5] J. Felsenstein. Phylogeny inference package (version 3.2). *Cladistics*, 5, 1989.

[6] O. Gascuel. BIONJ: An improved version of the NJ algorithm based on a simple model of sequence data. *Mol. Biol. Evol.*, 14:685–695, 1997.

[7] S. Kumar. A stepwise algorithm for finding minimum evolution trees. *Mol. Biol. Evol.*, 1996.

[8] T. Margush and F.R. McMorris. Consensus n-trees. *Bulletin of Math. Biol.*, 43(2):239–244, 1981.

[9] A. Rambaut and N.C. Grassly. Seq-gen: An application for the monte carlo simulation of dna sequence evolution along phylogenetic trees. *Comput. Appl. Biosci.*, 1997.

[10] M. Steel. The complexity of reconstructing trees from qualitative characters and subtrees. *J. of Classification*, 9:91–116, 1992.

[11] K. Strimmer, N. Goldman, and A. von Haeseler. Baysian probabilities and quartet puzzling. *Mol. Biol. Evol.*, 14:210–211, 1997.

[12] K. Strimmer and A. von Haeseler. Quartet puzzling: a quartet maximum-likelihood method for reconstructing tree topologies. *Mol. Biol. Evol.*, 13(7):964–969, 1996.

[13] S. J. Willson. Building phylogenetic trees from quartets by using local inconsistency measure. *Mol. Biol. Evol.*, 16:685–693, 1999.

Computational Complexity of Word Counting[*]

Mireille Régnier

INRIA, 78153 Le Chesnay

Abstract. Evaluation of the frequency of occurrences of a given set of patterns in a DNA sequence has numerous applications and has been extensively studied recently. We discuss the computational complexity for explicit formulae derived by several authors. We introduce a correlation automaton, that minimizes this complexity. This is crucial for practical applications. Notably, it allows to deal with the Markovian probability model. The case of patterns with some unspecified characters – approximate searching, regular expressions, ... – is addressed.

1 Introduction

Repeated patterns and related phenomena in sequences (also called words or strings) have been extensively studied recently, due to applications to molecular biology. The underlying assumption is that avoided or overrepresented words (may) have a specific biological function. One fundamental question that arises is the computation of the frequency of pattern occurrences in another string known as the *text*. Other parameters are studied, such as the waiting time for the first occurrence or the r-scans. These questions have been addressed for a (possibly infinite) set $\mathcal{H} = (H_i)_{1 \leq i \leq q}$ of patterns and various assumptions on the counting of possible overlaps. The text may be generated according either to the Bernoulli model or the Markovian model. Existing approaches either rely on probabilistic methods [2,16,20,1], such as the martingales, or algebraic methods: generating functions or recurrences [7,11,4,15,3,6,23,13,21], languages [19,17] or automata [14,22]. A first survey, from the point of view of applications to DNA sequences, can also be found in [12] with references to applications to computational biology.

Our goal is NOT to provide with new formulae or approaches. We rather point out what is the intrinsic complexity of word counting problems. This is crucial to practical applications: depending on the approach, the size (or the structure) of the pattern set and the probabilistic model imply limits to the tractability of formulae. In most approaches, it appears that the computation of the mean and the variance imply, at some step, the *inversion of a linear system*, or an equivalent operation such as the computation of a determinant, an infinite sum of matrices, etc. We briefly discuss when and why this step is unavoidable and then we study the minimal size of the linear system involved. This is precisely intrinsic complexity of the counting problems. We introduce the *correlation*

---

[*] This research was supported by ESPRIT LTR Project No. 20244 (ALCOM IT) and REMAG Action of INRIA.

O. Gascuel, M.-F. Sagot (Eds.): JOBIM 2000, LNCS 2066, pp. 99–110, 2001.

automaton and point out that its size provides the complexity. We will illustrate the gain in complexity with a few important classes of word counting considered in biological applications. Finally, we extend this discussion to study how much the introduction of a Markovian model increases the complexity defined for the Bernoulli model. One might expect that the computational complexity is multiplied by the size of the underlying Markov chain, which quickly leads to untractable formulae [22]. We will show on several typical examples that it is not so and that one can keep the computational complexity within reasonable limits.

2   Basic Tools

2.1   Probabilistic Models

The text generation follows either the Bernoulli model or the Markovian model of order K over an alphabet $\mathcal{S}$ of size V. One makes use of the transition matrix $\mathbb{P}$. Vector $\boldsymbol{\pi} = (\pi_1, \dots, \pi_{V^K})$ denotes the stationary distribution satisfying $\boldsymbol{\pi}\mathbb{P} = \boldsymbol{\pi}$ and $\boldsymbol{\Pi}$ is the stationary matrix that consists of V^K identical rows equal to $\boldsymbol{\pi}$. Finally, results depend on the fundamental matrix of the underlying Markov chain [8], defined as $\mathbb{Z} = (\mathbb{I} - (\mathbb{P} - \boldsymbol{\Pi}))^{-1}$ where $\mathbb{I}$ is the identity matrix. Here, $Z_{i,j} = \sum_n (\mathbb{P}^n_{i,j} - \boldsymbol{\Pi})$. To keep notations lighter, we restrict here to $K = 1$. In this special case, transition matrix is $\mathbb{P} = \{p_{i,j}\}_{i,j \in \mathcal{S}}$ where $p_{i,j} = \Pr\{t_{k+1} = j | t_k = i\}$. In the Bernoulli model, the rank of $\mathbb{P}$ is 1, e.g. $\mathbb{P} = \boldsymbol{\Pi}$. The probability to find a given character s from $\mathcal{S}$ is independent of previous characters and denoted p_s. Let $\mathcal{H}$ be a set of q patterns, $\{\mathrm{H}_i\}_{1 \le i \le q}$. Let $\mathbb{F}(z)$ be the $q \times q$ matrix defined by:

$$\mathbb{F}(z)_{i,j} = \frac{1}{\pi_{\mathrm{H}_j[1]}} [(\mathbb{P} - \varPi)(\mathbb{I} - (\mathbb{P} - \varPi)z)^{-1}]_{\mathrm{H}_i[m_i], \mathrm{H}_j[1]} \ ,$$

where $\mathrm{H}_j[1]$ denotes the first character of H_j and $\mathrm{H}_i[m_i]$ denotes the last character of H_i. Its *rank* is upper bounded by $\min(q, V^K)$. It is noteworthy that $\mathbb{F}(z)$ is the zero matrix in the Bernoulli model. In the Markovian model, coefficients of $\mathbb{F}(1)$ and $\mathbb{F}'(1)$ are extracted from $(\mathbb{P} - \varPi)\mathbb{Z}$ and $(\mathbb{P} - \varPi)^2 \mathbb{Z}^2$.

Below, $P(w_1|w_2)$ is the *conditional probability* that w_1 occurs at position k knowing that w_2 occurs at position $k - |w_2|$; $P(w)$ is the *stationary probability* that the word w occurs in the random text S between symbols k and $k + |w| - 1$ when k goes to infinity.

2.2   Overlap and Correlation Sets

It turns out that the pattern occurrences counts depend on the so-called correlation sets defined below. In algebraic approaches, one expresses this dependency through the so called *correlation polynomials* introduced in [7] for the Bernoulli model, extended in [19] to the Markov model.

Definition 1. *Given two strings H and F, the* overlap set *is the set of H-suffixes that are F-prefixes. F-suffixes of the associated F-factorizations form the* correlation set $\mathcal{A}_{H,F}$. *One defines the* **correlation polynomial** *of H and F as:*

$$A_{H,F}(z) = \sum_{w \in \mathcal{A}_{H,F}} P(w|H) z^{|w|}$$

When H is equal to F, $\mathcal{A}_{H,H}$ is named the autocorrelation set *and denoted $\mathcal{A}_H$; the* autocorrelation polynomial *is defined as:*

$$A_H(z) = \sum_{w \in \mathcal{A}_H} P(w|H) z^{|w|} \ .$$

Intuitively, when a word in $\mathcal{A}_{H,F}$ is concatenated to H, it creates an (overlapping) occurrence of F. For example, let $H = 11011$ and $F = 1110$. Then $\mathcal{A}_{H,F} = \{10, 110\}$ and $\mathcal{A}_{F,H} = \{11\}$. The associated correlation polynomials are, in a biased Bernoulli model where $(p_0, p_1) = (1/3, 2/3)$, $A_{11011,1110}(z) = \frac{2}{9}z^2 + \frac{4}{27}z^3$ while $A_{1110,11011}(z) = \frac{4}{9}z^2$. The autocorrelation polynomials are: $A_{1110}(z) = 1$ and $A_{11011}(z) = 1 + \frac{4}{27}z^3 + \frac{8}{81}z^4$.

Assume now that $H = CGC$ over alphabet $\mathcal{S} = \{A, C, G, T\}$. Observe that $\mathcal{A}_{H,H} = \{\epsilon, GC\}$, where ϵ is the empty word. Thus, for the uniform Bernoulli model (all symbols occur with the same probability equal to 0.25), $A_{CGC}(z) = 1 + \frac{z^2}{16}$, while for the Markovian model of order one, $A_{CGC}(z) = 1 + p_{C,G}p_{G,C}z^2$.

Remark 1. For a Markov model of order $K = 1$, $A_H(z)$ and $A_{H,F}(z)$ only depend on the last character of H.

Definition 2. *Let $\mathcal{H}$ be a given set of q searched patterns, denoted $(H_i)_{i=1...q}$. Correlation matrix is the $q \times q$ matrix of correlation polynomials:*

$$\mathbb{A}(z) = ||A_{H_i,H_j}(z)||_{i,j=1,q} \ .$$

Let $\mathbf{H}(z)$ be the row vector $(P(H_1)z^{m_1}, \ldots, P(H_q)z^{m_q})$. Let $\mathbb{H}$ the $q \times q$ matrix with q identical rows that are equal to $\mathbf{H}(z)$.

3   Word Counting Complexity

Various counting models may be considered. We restrict here our discussion to the two most common ones, namely the *overlapping model* and the *renewal model*. In the overlapping model, any occurrence of a pattern in a text is counted. In the renewal model, an occurrence is counted iff it does not overlap on the left with an other occurrence that is already counted. For example, consider text $TTATATATG$ and pattern TAT. One counts three occurrences of TAT in the overlapping model, and only two in the renewal model (at positions 2 and 6).

It is proven, partially or totally, by various methods, in the works referred above that:

Theorem 1. *Let $\mathcal{H}$ be a set of patterns H_i of sizes m_i. Let $X_{i,n}$ be the random variable that counts H_i-occurrences in a text of size n. In the Markov and Bernoulli case, in the renewal model, the row vector of expectations, $(E(X_{1,n}), \cdots, E(X_{q,n}))$, is:*

$$n\mathbf{H}(1)\mathbb{A}(1)^{-1} + [\mathbf{H}(1)\mathbb{A}(1)^{-1} + \mathbf{H}(1)\mathbb{A}(1)^{-1}\mathbb{A}'(1)\mathbb{A}(1)^{-1} - \mathbf{H}'(1)\mathbb{A}(1)^{-1}] \ . \quad (1)$$

As these results are explicit formulae. They are *exact* in the Bernoulli model. In the Markovian model, the approximation order is exponentially small. It follows that, in the general case, the computation of the mean in the renewal model involves, for both probabilistic models, the inversion of $\mathbb{A}(1)$, which is a matrix of size $|\mathcal{H}| \times |\mathcal{H}|$.

It is worth giving here an intuition for the appearance of term $\mathbb{A}(1)^{-1}$. In the renewal model, a rather subtle dependance to the past appears. Namely, the validity of an occurrence at position i depends on the chain of overlapping occurrences ending at position i, if any. From definitions above, it follows that $P(H_i) \cdot [A_{H_i,H_j}(1) - 1]$ represents the probability for an overlapping occurrence of H_i and H_j ending at some given position k. Hence, the row vector $\mathbf{H}(1)(\sum_{n \geq 1}(\mathbb{A}(1) - \mathbb{I})^n)$ represents the cumulated probability of all finite overlapping chains ending with $H_1, \cdots, H_q$. This is precisely $\mathbf{H}(1)\mathbb{A}(1)^{-1}$.

Let us now turn to *higher moments*. Theorem below is proved in [17], and partial results can be found elsewhere. Notably, the Bernoulli model is covered in [2,21,19]. Partial results in the Markov case can be found in [13,23].

In both counting models, the variance and covariance only depend on the probabilities of the sets and the correlation sets. Nevertheless, the dependency is different (as the means are different). This leads to the formal definitions below:

Definition 3. *Let $\mathcal{H}$ be a set of patterns H_i of sizes m_i. Let:*

(i)   Overlapping model:

$$\mathbb{L}(z) = I\!H(z) \ ,$$
$$\mathbb{C}(z) = I\!H(z).(\mathbb{A}(z) - \mathbb{I})$$

(ii)   Renewal model:

$$\mathbb{L}(z) = I\!H(z)\mathbb{A}(z)^{-1} \ ,$$
$$\mathbb{C}(z) = 0 \ .$$

Define in both models:

$$\mathbb{B}(z) = \mathbb{L}(z).\mathbb{L}(z)^t \ ,$$
$$I\!E(z) = \mathbb{L}(z).(\mathbb{F}(z)\mathbb{L}(z))^t \ .$$

Theorem 2. *Let* $\mathcal{H}$ *be a set of patterns* H_i *of sizes* m_i. *The variance-covariance matrix is equal to:*

$$n[\mathbf{X}_1 + \mathbf{X}_2] + [\mathbf{Y}_1 + \mathbf{Y}_2] \ , \tag{2}$$

where:

$$\begin{aligned}
\mathbf{X}_1 &= \mathbb{B}(1) - \mathbb{B}'(1) + \mathbb{C}(1) + Diag(\mathbb{L}(1)) \\
\mathbf{Y}_1 &= \mathbf{X}_1 + \mathbb{B}''(1) - \mathbb{L}'(1).\mathbb{L}'(1)^t - \mathbb{C}'(1) - Diag(\mathbb{L}'(1)) \\
\mathbf{X}_2 &= \mathbb{E}(1) + \mathbb{E}(1)^t \\
\mathbf{Y}_2 &= \mathbf{X}_2 - (\mathbb{E}'(1) + \mathbb{E}'(1)^t) \ ,
\end{aligned}$$

In both cases, $\mathbf{X}_2$ *and* $\mathbf{Y}_2$ *reduce to* 0 *in the Bernoulli model.*

In the Bernoulli probabilistic model, and overlap counting, it appears that *no inversion* is necessary, even when $\mathcal{H}$ is not a singleton. In other words, the linear systems that appear in recurrence approaches such as [7,5] or searching automaton approach [14] are unnecessary steps.

Here, the dependency to Markov model comes from $\mathbb{E}(1)$ and $\mathbb{E}'(1)$, which represent the fundamental matrix $\mathbb{Z}$ of the underlying Markov chain and $\mathbb{Z}^2$. It follows that, as long as we are only concerned with finite moments, we have two *independent systems* to invert. The computational complexity is *not the product* $|\mathcal{H}| \times V^K$ but the *sum* $|\mathcal{H}| + V^K$.

Clearly, word counting does not reduce to word occurrence counting. The waiting time for the first occurrence, texts with at least one occurrence of a given word or a given set of words [7,5] are of interest. Nevertheless, general scheme developped in [19,17] shows that the general flavor of this discussion is still correct.

It appears that *distribution results* are essential to many applications [9]. Let $P(O_n(\mathcal{H}) \geq r)$ be the probability that at least r occurrences of $\mathcal{H}$-patterns occur in a random text of size n. Probabilistic approaches [2,16] prove that limiting distribution is Gaussian. We introduced in [19] the bivariate generating function $T(z, u)$:

$$T(z, u) = \sum_{n, r \geq 0} P(O_n(\mathcal{H}) \geq r) u^r z^n$$

that provides formulae to compute probabilities in the finite range that are stable numerically [9]. A linear equation for $T(z, u)$ is derived in [17], under general constraints on set $\mathcal{H}$ that only depend on correlation matrix $\mathbb{A}(z)$, fundamental matrix $\mathbb{Z}$ (through $\mathbb{F}(z)$) and $\mathbf{H}(z)$.

4   Correlation Automaton

We mentioned above that all results on $\mathcal{H}$-occurrences depend on the possible overlaps of patterns in set $\mathcal{H}$. If this set is large, computing explicit formulae

[2,21,17] involves the computation of an autocorrelation matrix of size $|\mathcal{H}| \times |\mathcal{H}|$; brute force may lead to untractable formulae.

A key observation is the fact that not all patterns overlap with all other patterns. I.e. the matrix is sparse. In the case where the counting of each pattern separately is actually necessary, one should rely on this to derive the computation efficiently. We will not discuss this point here. Additionnally, one can use the sparsity of transition matrix [22].

We rather point out that, in the case where one counts all possible occurrences within the set $\mathcal{H}$ simultaneously, it is possible to *aggregate the patterns*. As a matter of fact, writing a set as a (non-ambiguous) regular expression, is a first aggregation of the set of possible instantiations. In [14], an adaptation of algorithms searching for regular expressions allows for the computation of mean and variance for any given regular expression. Nevertheless, this approach does not provide explicit formulae in terms of overlapping polynomials. The size of the linear system to be inversed is the size of the *searching automaton*, which leads to severe limitations on the sets that are tractable.

We introduce a new concept, the *stable partition*, and a new automaton, the *autocorrelation automaton*.

Definition 4. *Let $\mathcal{H}$ be a set of patterns and $(\mathcal{H}_i)_{1 \leq i \leq q}$ be a finite partition. Partition $(\mathcal{H}_i)_{1 \leq i \leq q}$ is stable iff it satisfies the following property:*

$$\forall (i,j), \ \exists A_{ij} \ s.t. : \forall H_i \in \mathcal{H}_i : \mathcal{A}_{i,j} = \cup_{H_j \in \mathcal{H}_j} \mathcal{A}_{H_i, H_j} \ . \tag{3}$$

Fact 1: $\mathcal{H}$ satisfies Property (3) iff

$$\forall (i,j) \cup_{H_i \in \mathcal{H}_i, H_j \in \mathcal{H}_j} \mathcal{A}_{H_i, H_j} = \cap_{H_i \in \mathcal{H}_i} \cup_{H_j \in \mathcal{H}_j} \mathcal{A}_{H_i, H_j} \ .$$

Example 1. Let $\mathcal{H}$ be the PROSITE expression $[ST]x(2)[DE]$. Here [ ] stands for a choice and ( ) for a length. One has $|\mathcal{H}| = 2^2.20^2 = 1600$. Partition:

$$\mathcal{H}_1 = \{[ST][\bar{S}T][ST][DE]\} \ , \ \mathcal{H}_2 = \{[ST](2)[\bar{S}T][DE]\},$$
$$\mathcal{H}_3 = \{[ST](3)[DE]\} \ , \ \mathcal{H}_4 = \mathcal{H} - (\mathcal{H}_1 \cup \mathcal{H}_2 \cup \mathcal{H}_3)$$

defines a stable partition of $\mathcal{H}$. The non-empty correlation sets are:

$$\mathcal{A}_{1,1} = \{[ST][DE]\}, \ \mathcal{A}_{1,4} = \{[\bar{S}T][DE]\}, \ \mathcal{A}_{2,4} = \{[DE]\} \ ,$$
$$\mathcal{A}_{3,2} = \{[DE]\}, \ \mathcal{A}_{3,1} = \{[ST][DE]\}, \ \mathcal{A}_{3,4} = \{[\bar{S}T][DE]\}$$

In multiset counting, one assumes [2,19,21,14] that the set is *reduced*, e.g. that no H_i is a suffix of any H_j. This property does not transfer automatically to a stable partition but is guaranteed below by the introduction of $\bar{\mathcal{A}}_{i,j}$.

Definition 5. *Let $\mathcal{H}$ be a set of patterns and $(\mathcal{H}_i)_{1 \leq i \leq q}$ be a finite partition satisfying Property (3). The correlation automaton is the graph with q vertices labelled by sets $\mathcal{H}_i$ and edges labelled by $\tilde{\mathcal{A}}_{i,j}$, where*

$$\tilde{\mathcal{A}}_{i,j} = \{w \in A_{ij}; w \text{ has no proper prefix in any } A_{i,k}\}$$

The correlation automaton associated to set $\mathcal{H}$ above is depicted below:

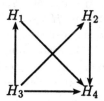

In the example above, one gets $\tilde{\mathcal{A}}_{3,4} = \{[ST\bar{D}E][DE]\}$ that is different from $\mathcal{A}_{3,4}$. In other words, the concatenation of a word in $[DE][DE]$ right of an instantiation of $\mathcal{H}_3$ creates an instantiation of $\mathcal{H}_2$ left of an instantiation of $\mathcal{H}_4$. Hence, these words are not in $\tilde{\mathcal{A}}_{3,4}$.

Definition 6. *Assume a Bernoulli model. With the definitions above, one defines:*

$$A_{i,j}(z) = \sum_{w \in \tilde{\mathcal{A}}_{i,j}} P(w)z^{|w|} \ .$$

The correlation matrix of set $\mathcal{H}$ with stable partition $(\mathcal{H}_i)_{1 \leq i \leq q}$ is the $q \times q$ matrix:

$$\mathbb{A}(z) = ||A_{i,j}(z)||_{1 \leq i \leq q}$$

Theorem 3. *Given a set $\mathcal{H}$ and a partitioning satisfying the property above, the mean and the variance of the expected number of $\mathcal{H}$-occurrences are linear functions of the size of the text. The linearity constants derived in [17] hold. These results are exact for the Bernoulli model, and the approximation is exponentially decreasing for the Markovian model.*

The key observation is that all equations written for singletons in [17] extend for sets $\mathcal{H}_i$ that satisfy Property (3). Moreover, the computation of $A_{i,j}(z)$ reduces to a summation: $(\sum P(w)z^{|w|}$ in the Bernoulli case). This reduces the complexity of the derivation of the expected number of occurrences, as well as the variance: it depends on the *size of the partition*.

5   Various Examples

We discuss the size of the correlation automaton on some examples.

5.1   1-Partitions

This simple case appears to be rather interesting. First, any motif set $\mathcal{H}$ of size 1 trivially defines a 1-partition, $\mathcal{H}$ itself. This case has been extensively studied and the Markovian case is considered in [5,19,20]. It is proven [19] that bivariate generating function $T(z, u)$ satisfies *one linear equation*. It only depends of the autocorrelation polynomial and some coefficients of matrix $\mathbb{F}(z)$. Mean and variance are known to be linear functions of the size n of the text and these functions are given in [19].

A second important case occurs when $\mathcal{H}$ is a set of non-overlapping patterns. $\mathcal{H}$ is a 1-partition that satisfies the property above, with $\mathcal{A}_{\mathcal{H},\mathcal{H}} = \emptyset$, and all results provided in [19] for a single pattern steadily apply in the Bernoulli and the Markovian case. The reason is that the correlation matrix is equal to identity when patterns are non overlapping. Alternatively, this follows from a simplification of our general result in Theorem 1 and Theorem 2.

As a conclusion, when $\mathcal{H}$ has a stable 1-partition, no system inversion is necessary and one only computes, in the Markov case, a matrix with *real coefficients*, $\mathbb{Z}$. This is a large improvement on initial results in [5] where a $3 \times V^K$ linear system is solved, where variables are *formal series*.

Indeed, we have a similar simplification when we relax the non-overlapping constraint by allowing each pattern to overlap with itself only. In that case, the correlation matrix $\mathbb{A}(z)$ is diagonal.

5.2   Approximate Matching

It appears of particular interest to be able to evaluate the expected number of *approximate* occurrences of a given word H_0 [10]. In [22], M. Tompa aggregates the states of the searching automaton, and obtains $1.5 \times |H_0|^2$ states when $V = 4, K = 1$. Let us discuss on a specific example the size of the correlation automaton and the additional complexity due to Markovian hypotheses. Let H_0 be the pattern *abacaba* and $\mathcal{H}$ the patterns which are within distance 1 according to the Hamming distance (at most one substitution is allowed). Here, $|H_0| = 7$ and $|\mathcal{H}| = V \times H_0$. $\mathcal{H}$ can be partitioned into 4 sets, in the Bernoulli case:

$$\mathcal{F}_1 = \{abaxaba; x \neq c\} \quad ; \quad \mathcal{F}_2 = \{wcaba; w \neq aba, |w| = 3\}$$
$$\mathcal{F}_3 = \{abacabx; x \neq a\} \quad ; \quad \mathcal{F}_4 = \{abacwa; w \neq ab, |w| = 2\}$$

As the last character of $\mathcal{F}_3$ is not a, a $\mathcal{F}_3$-pattern can only overlap with an $\mathcal{F}_2$-pattern, and $\tilde{\mathcal{A}}_{3,2} = \{bacaba\}$. Any other subset can overlap with any subset $\mathcal{F}_i$ of the partition:

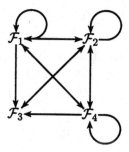

Typical examples for edges are $\tilde{A}_{1,1} = \{\bar{c}aba, ba\{c,\bar{b}\}aba\}$; $\tilde{A}_{4,2} = \{caba, \bar{b}acaba\}$. This example shows that minimal correlation automaton size depends on H_0-structure, more precisely on its periods. One expects the size to be linearly bounded by $|H_0|$ and V^K for "non-pathological" structures of H_0. Finally, observe that final states $\mathcal{F}_i$ need not be split in the Markovian model. Indeed, one only needs to split some edges, to compute the correlation matrix, which essentially involves addition-multiplications. An efficient algorithm to compute the mean and the variance in that case is described in [18].

5.3   PROSITE Expressions

The correlation automaton allows for a huge improvement when regular expressions are considered [14]. For example, let $\mathcal{H}$ be the set of patterns that instantiate PROSITE expression PS00844,i.e.:

$$[LIVM]x(3)[GA]x[GSAIV]R[LIVCA]D[LIVMF](2)x(7,9)$$
$$[LI]xE[LIVA]N[STP]xP[GA] \ .$$

The cardinality of this set is about 1.9×10^{26} and the searching automaton has 946 states in the Bernoulli case [14]. Nevertheless, this set $\mathcal{H}$ can be partitioned into 5 states. This rather surprising fact can be explained by the fact that, despite the number of unspecified choices, only a few overlaps are allowed. Namely, let:

$$\mathcal{F}_1 = [LIVM]x(3)[GA]x[GSAIV]R[LIVCA]D[LIVMF](2)x(7,9)$$
$$[LI]xE[LIVA]N[STP][LIVM]P[GA]$$
$$\mathcal{F}_2 = [LIVM]x(3)[GA]x[GSAIV]R[LIVCA]D$$
$$[LIVM][LIVMF]x(2)[GA]x[GSAIV]R[LIVCA]D$$
$$[LI][LIVMF]E[LIVA]N[STP][LI\bar{V}M]P[GA]$$

$$\mathcal{F}_3 = [LIVM]x(3)[GA]x[GSAIV]R[LIVCA]D$$
$$[LIVM][LIVMF]x(2)[GA]x[GSAIV]R[LIVCA]D$$
$$[LIVMF][LI]xE[LIVA]N[STP][LI\bar{V}M]P[GA]$$
$$\mathcal{F}_4 = [LIVM]x(3)[GA]x[GSAIV]R[LIVCA]D[LIVMF]$$
$$[LIVM]x(3)[GA]x[GSAIV]R[LIVCA]D[LI]$$
$$[LIVMF]E[LIVA]N[STP][LI\bar{V}M]P[GA]$$
$$\mathcal{F}_5 = \mathcal{H} - (\mathcal{F}_1 \cup \mathcal{F}_2 \cup \mathcal{F}_3 \cup \mathcal{F}_4)$$

Any pattern in $\mathcal{F}_1$ can overlap on the right with a new instantiation, as its 3-suffix is an allowed 3-prefix. There are only three other final sets that allow overlap. The sets $\mathcal{F}_2, \mathcal{F}_3$ and $\mathcal{F}_4$ are subsets of $\mathcal{H}$ where the number of unspecified characters between $[LIVMF](2)$ and $[LI]$ are respectively $8, 9$ and 9. The allowed shifts are respectively $10, 10$ and 11. No other shift is allowed. It steadily follows that an $\mathcal{H}$-pattern in $\mathcal{F}_2, \mathcal{F}_3$ or $\mathcal{F}_4$ cannot be in $\mathcal{F}_2, \mathcal{F}_3$ or $\mathcal{F}_4$: otherwise, shifts 20 or/and 21 would be allowed. In other words, these three subsets only overlap with $\mathcal{F}_1$ and $\mathcal{F}_5$. By definition, $\mathcal{F}_5$ cannot overlap on the right with any $\mathcal{H}$-pattern. All these properties can be summarized in the correlation automaton depicted below.

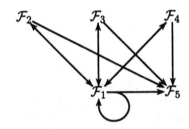

Moreover, considering a Markovian model does not multiply computational complexity by V^K. As a matter of fact, all instantiations of $\mathcal{H}$ end with the same suffix $[P][GA]$. If $K \leq 2$, the computational complexity of the Markovian and Bernoulli models are equivalent... This problem will be detailed in a forthcoming paper, that will include algorithmic issues -how to build correlation automaton- that are beyond the scope of this paper. An algorithm that computes the expectation, using dynamic programming, is presented in [18].

6   Conclusion

We briefly showed how different approaches such as *languages and regular expression searching automata* can be combined to provide an efficient computation of the distribution of pattern occurrences from a given set. We introduced a *correlation automaton* and showed how it allows for a drastic decrease of computational complexity in word counting, notably when text is generated according to a Markovian model. One minimizes the complexity, e.g. the rank of some linear system or matrix. Additionnally, the derivation of mean and variance reduce to computations with real numbers -some underlying probabilities-, a real advantage on computations that deal with formal variables. We are currently working to provide efficient algorithms that compute correlation automaton in various cases of particular interest: approximate occurrences of a word, regular expressions that characterize a family of sequences,... This has numerous applications in computational biology, where current implementations suffer from severe drawbacks due to the explosion of computational complexity.

References

1. Apostolico, A., Bock, M., Lonardi, S., and Xu, X. (1999). Efficient detection of unusual words. *Journal of Computational Biology*. to appear; preliminary version as Technical Report 97-050, Purdue University Computer Science Department (1996).
2. Bender, E. A. and Kochman, F. (1993). The Distribution of Subwords Counts is Usually Normal. *European Journal of Combinatorics*, 14:265–275.
3. Borodovsky, M. and Kleffe, J. (1992). First and second moments of counts of words in random texts generated by markov chains. *CABIOS*, 8:433–441.
4. Breen, S., Waterman, M., and Zhang, N. (1985). Renewal theory for several patterns. *J. Appl. Prob.*, 22:228–234.
5. Chrysaphinou, C. and Papastavridis, S. (1990). The occurrence of sequence of patterns in repeated dependent experiments. *Theory of Probability and Applications*, 79:167–173.
6. Geske, M., Godbole, A., Schafner, A., Skolnick, A., and Wallstrom, G. (1995). Compound Poisson Approximations for Word Patterns Under Markovian Hypotheses. *J. Appl. Prob.*, 32:877–892.
7. Guibas, L. and Odlyzko, A. (1981). String Overlaps, Pattern Matching and Nontransitive Games. *Journal of Combinatorial Theory,* Series A, 30:183–208.
8. Kemeny, J. and Snell, J. (1983). *Finite Markov Chains*. Undergraduate Texts in Mathematics. Springer-Verlag, Berlin.
9. Klaerr-Blanchard, M., Chiapello, H., and Coward, E. (2000). Detecting localized repeats in genomic sequences: A new strategy and its application to *B. subtilis* and *A. thaliana* sequences. *Comput. Chem.*, 24(1):57–70.
10. Kurtz, S. and Myers, G. (1997). Estimating the Probability of Approximate Matches. In *CPM'97*, Lecture Notes in Computer Science. Springer-Verlag.
11. Li, S. (1980). A Martingale Approach to the Study of Occurrences of Sequence Patterns in Repeated Experiments. *Ann. Prob.*, 8:1171–1176.
12. Li, W. (1997). The study of correlation structures of DNA sequences: a critical review. *Computers Chem.*, 21(4):257–271.
13. Lundstrom, R. (1990). *Stochastic Models and Statistical Methods for DNA Sequence Data*. Phdthesis, University of Utah.
14. Nicodème, P., Salvy, B., and Flajolet, P. (1999). Motif statistics. In *ESA '99*, volume 1643 of *Lecture Notes in Computer Science*, pages 194–211. Springer-Verlag. Proc. European Symposium on Algorithms-ESA'99, Prague.
15. Pevzner, P., Borodovski, M., and Mironov, A. (1991). Linguistic of Nucleotide sequences:The Significance of Deviations from the Mean: Statistical Characteristics and Prediction of the Frequency of Occurrences of Words. *J. Biomol. Struct. Dynam.*, 6:1013–1026.

16. Prum, B., Rodolphe, F., and de Turckheim, E. (1995). Finding Words with Unexpected Frequencies in DNA sequences. *J. R. Statist. Soc. B.*, 57:205–220.

17. Régnier, M. (2000). A Unified Approach to Word Occurrences Probabilities. *Discrete Applied Mathematics*, 104(1):259–280. Special issue on Computational Biology;preliminary version at RECOMB'98.

18. Régnier, M., Lifanov, A., and Makeev, V. (2000). Three variations on word counting. In *GCB'00*, pages 75–82. Logos-Verlag. Proc. German Conference on Bioinformatics, Heidelberg.

19. Régnier, M. and Szpankowski, W. (1997). On Pattern Frequency Occurrences in a Markovian Sequence. *Algorithmica*, 22(4):631–649. preliminary draft at ISIT'97.

20. Schbath, S. (1995). *Etude Asymptotique du Nombre d'Occurrences d'un mot dans une Chaine de Markov et Application à la Recherche de Mots de Frequence Exceptionnelle dans les Sequences d'ADN*. Thèse de 3e cycle, Université de Paris V.

21. Tanushev, M. and Arratia, R. (1997). Central Limit Theorem for Renewal Theory for Several Patterns. *Journal of Computational Biology*, 4(1):35–44.

22. Tompa, M. (1999). An exact method for finding short motifs in sequences, with application to the ribosome binding site problem. In *ISMB'99*, pages 262–271. AAAI Press. Seventh International Conference on Intelligent Systems for Molecular Biology, Heidelberg,Germany.

23. Waterman, M. (1995). *Introduction to Computational Biology*. Chapman and Hall, London.

EuGène: An Eukaryotic Gene Finder That Combines Several Sources of Evidence

Thomas Schiex[1], Annick Moisan[1], and Pierre Rouzé[2]

[1] INRA, Toulouse, France,
Thomas.Schiex@toulouse.inra.fr,
http://www-bia.inra.fr/T/schiex
[2] INRA, Gand, Belgique
Pierre.Rouze@gengenp.rug.ac.be

Abstract. In this paper, we describe the basis of EuGène, a gene finder for eukaryotic organisms applied to *Arabidopsis thaliana*. The specificity of EuGène, compared to existing gene finding software, is that EuGène has been designed to combine the output of several information sources, including output of other software or user information. To achieve this, a weighted directed acyclic graph (DAG) is built in such a way that a shortest feasible path in this graph represents the most likely gene structure of the underlying DNA sequence.

The usual simple Bellman linear time shortest path algorithm for DAG has been replaced by a shortest path with constraints algorithm. The constraints express minimum length of introns or intergenic regions. The specificity of the constraints leads to an algorithm which is still linear both in time and space.

EuGène effectiveness has been assessed on Araset, a recent dataset of *Arabidopsis thaliana* sequences used to evaluate several existing gene finding software. It appears that, despite its simplicity, EuGène gives results which compare very favourably to existing software. We try to analyse the reasons of these results.

Motivations

It is standard, in a thorough sequence annotation, to take into account several sources of evidence in order to try to precisely locate genes (exons/introns) in eukaryotic sequences. The sources exploited typically include:

- matches against databases (cDNA, EST, protein databases...);
- output of splice sites or translation start prediction software;
- more or less sophisticated "integrated" gene finding software, eg. GeneMark.hmm [9].
- experimental evidence or human expertise.

None of these sources of evidence is, alone, sufficient to decide gene locations and the manual integration of all these data is a painful and extremely slow work. The motivation of our work is, as far as possible, to automate this job using

O. Gascuel, M.-F. Sagot (Eds.): JOBIM 2000, LNCS 2066, pp. 111–125, 2001.

Arabidopsis thaliana as a first test organism. Several integrated gene finders exist that integrate protein or cDNA homologies in their prediction [16,8,6]. EuGène is naturally closely related to these tools. It most striking peculiarity is it's parasitic behaviour: EuGène has been designed to exploit other tools or information sources, including human expertise or other integrated gene finding programs. This allows to easily select a combination of ingredients..

1   Methods

To be able to integrate basic information on a genomic sequence, we have used a simple, general, efficient and yet effective graph-based approach for gene finding that allows to combine several sources of evidence. Rather than directly combining the output of existing gene finding software (as in [10]) we decided to combine the information at the lowest level in order to be able to:

1. maintain the consistency of the prediction;
2. globally assess the impact of each local choice w.r.t. all available evidence.

Given a raw DNA sequence, the basic idea is to consider a directed acyclic graph (DAG) such that all possible consistent gene structures are represented by a path in the graph. The gene structure currently used in EuGène is the simplest reasonable structure, the only signals taken into account being translation initiation sites (`ATG`), translation termination sites (stops) and splice sites. It can handle multiple genes which may be partial on the extremities of the sequence and explores the two strands of the sequence simultaneously as it is now standard in gene finding.

The DAG used for a simple ad-hoc sequence "`CATGAGGTAGTGA`" is illustrated in Fig. 1. It is a graph with 13 different tracks that correspond respectively to the 6 forward/reverse coding frames, 6 forward/reverse intronic phases and a so-called intergenic track that covers both true intergenic regions and transcribed untranslated regions (UTR). Each signal occurrence, between two successive nucleotides, generates one or more "switches" between two parallel tracks. Each source-sink path defines a sequence of consistent genes structures. The size of the graph is in $O(n)$ where n is the sequence length.

To choose one path among the $O(13^n)$ paths in this graph, each edge e is weighted by a positive number w_e in such a way that shortest paths in the graph correspond to gene structures that "best respect" the available evidence. A probabilistic interpretation of this model can be given as follows: each edge e has a probability of existence P_e. Under simple independence assumptions, the reliability of a source-sink path is simply defined by the product of all the P_e in the path. A most reliable path is then a shortest path in the graph where $w_e = -\log(P_e)$. The approach is comparable (although not equivalent) to HMM with a non-homogeneous transition matrix between hidden states (our tracks).

Given a directed acyclic graph, the simplest linear time, linear space shortest path algorithm is Bellman's algorithm [1]. It can output an optimal pos-

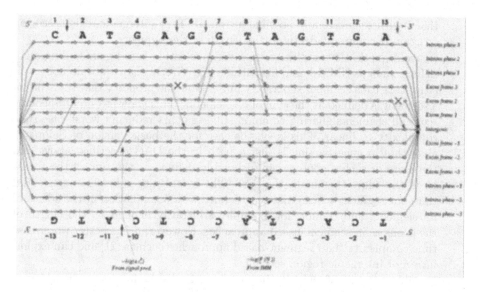

Fig. 1. The DAG explored by EuGène for a simple sequence. ATG occur at position 2 and −11, Stop (TGA at position 6 and 14, a donor at position 7, an acceptor at position 9.

sible gene structure[1]. Obviously, the use of Bellman's algorithm on the above DAG is similar to the approach of several existing gene finding algorithms. Like all recent gene finding algorithms, it does in-frame scoring and assembly [21]. More specifically, gene finding algorithms can be classified as segment-based or nucleotide-based:

- Segment-based approaches implicitly or explicitly build an exhaustive list of potential exons and/or introns which they separately score. This is the case for the algorithm presented in [21] or the algorithm presented in [13]. Note that in the worst case, the number of possible consistently signal-delimited segments grows quadratically in the length of the sequence. Since each potential segment must be scored, such segment-based approaches [21] are (at least) quadratic in the length of the sequence. The strength of these algorithms is that they allow essentially arbitrary segment scoring functions. To improve expected time complexity, one can exploit the property that exonsegments cannot contains in frame stop codons: the assumption that in frame stop codons occurs following a Poisson process suffices to make the expected number of possible exons linear in the length of the sequence. The algorithm presented in [13] in some sense exploits this property by only taking into account exons scores in the global score: it is worst-case linear in the number of exons and has therefore an expected linear complexity in the length of the sequence (under the Poisson assumption). The essential limitation is

---

[1] Linear space is needed to recover the shortest paths found.

that it either ignores any possible intron, UTR or intergenic regions scoring or again becomes quadratic in the length of the sequence.

– to avoid quadratic complexity, so-called nucleotide-based gene finding algorithms, such as HMM-based algorithms (eg. Viterbi algorithm) do not explicitly score each segment separately but exploit the property that the score of a segment is defined as the sum of its elements. This allows to simultaneously score all segments sharing common sub-segments reducing worst-case time complexity to linear. This is eg. the case for the basic version of Bellman's algorithm: a gene parsing is just a path in the DAG described above and the score of such a gene parsing is simply defined as the sum of the scores of all the edges that make the path. The complexity is linear in space and time but segment scoring must be defined as a combination of the scores of its elements and cannot, for example, arbitrarily depend on the length of the exons as in segment-based approaches. However, it has better worst-case time complexity than segment-based approaches such as [21] and can exploit intron or intergenic region scoring.

Possible variants of the Viterbi algorithm that take into account arbitrary explicit probability distributions on state lengths do exist [15] and have effectively been used in gene finding [3], but they become worst-case time quadratic. EuGène's algorithm is a sophistication of Bellman's algorithm that try to optimise the compromise between efficiency and generality: it is linear both in time and space but allows to take into account constraints on the minimum length of some gene elements (introns, single exon genes, intergenic regions). Basically, this means that the score of a segment is effectively the sum of the score of its elements unless the segment is too short: in this case the score becomes $-\infty$. The approach is comparable to explicit state duration HMM with uniform duration densities (see [14], pp. 270). Maximum length constraints can also probably be taken into account while keeping linear time and space but we haven't find any practically significant use of such constraints in gene finding.

The complete algorithm actually used in EuGène allows to specify minimum length constraints for intergenic regions depending on the orientation of the genes that border the region: convergent ($\rightarrow\leftarrow$), divergent ($\leftarrow\rightarrow$) or confluent ($\rightarrow\rightarrow$ or $\leftarrow\leftarrow$) genes. In the sequel, for the sake of simplicity, we only describe the version which takes into account minimum length constraints independently of the orientation of surrounding elements[2].

1.1   Algorithm

Given a nucleic acid sequence of length n, we orient it so that the 5′ end is on the left and the 3′ end on the right. Naturally, the reverse strand is oriented in the

---

[2] The complete version is not essentially different, remains linear, but requires the introduction of additional tracks in the graph. Note that merged stops may also occur in the predictions of the algorithm presented. . Again, this can be easily dealt with using additional tracks.

reverse direction. Nucleotides are numbered both from 1 to n (forward strand) and from -1 to $-n$ (reverse strand). We denote sites by their position and their type. The position of a site is an integer (positive or negative but different from 0) such that site p occurs before nucleotide p. The type t of a site can be either (I) for a translation initiation site, (A) for an acceptor site, (D) for a donor site or (T) for a termination site. For any given site, the frame of a site (p, t) is noted $F(p)$ and is defined as $sign(p).((((|p|-1) \bmod 3)+1)$. As exemplified in Fig. 1, we consider that both initiation (ATG) and termination sites are part of their exons which means that initiation sites occurs on the left of the ATG but termination sites occur on the right of the stop codons.

Given a sequence of length n and a list of sites (p, t) defined by their position p and their type t, we can define the directed acyclic graph $G = (V, E)$ explored by EuGène. The set of vertices V contains $2 + 13 \times (2n + 2)$ vertices. There is a source vertex s and target vertex t. For any given nucleotide at position $1 \le i \le n$ in the sequence, there are 13×2 vertices. The first 13 vertices are called the left vertices of nucleotide i and are denoted by $v_{i,j}^l$, $-6 \le j \le 6$. The 13 remaining ones are the right vertices and are denoted by $v_{i,j}^r$, $-6 \le j \le 6$. Furthermore, there are 13 extra right vertices before the first nucleotide which are denoted by $v_{0,j}^r$ and 13 extra left vertices after the last nucleotide denoted by $v_{n+1,j}^l$. For all these vertices, j is called the track of the vertex.

For a given nucleotide, from a modelling point of view, the 13 (right/left) vertices correspond to different states or tracks:

- track 0 correspond to intergenic (i.e., UTR and real intergenic) predictions
- tracks 1 to 3 (resp. -1 to -3) correspond to exons in frame 1 to 3 (resp. -1 to -3)
- tracks 4 to 6 (resp. -4 to -6) correspond to introns in phase 1 to 3 (resp. -1 to -3). The phase of an intron is the position were the splicing occurs wrt. nucleotides: phase 1 introns splice at the frontier of two codons, phase 2 introns splice before the 2nd base of codons and phase 3 introns splice before the 3rd base of codons.

The set of edges E contains:

- edges $(s, v_{0,j}^r)$, for all j, $-6 \le j \le 6$. These are the so-called "prior" edges.
- edges $(v_{n+1,j}^l, t)$, for all j, $-6 \le j \le 6$. These are the so-called "posterior" edges.
- edges $(v_{i,j}^l, v_{i,j}^r)$, for all positions i and for all j, $-6 \le j \le 6$. These are the "content" edges.
- edges $(v_{i,j}^r, v_{i+1,j}^l)$, for all positions i and for all j, $-6 \le j \le 6$. These edges are the "no site" edges.
- Then, for each site (p, t) occurring on the forward strand $(p > 0)$, extra edges are added according to t. These edge are called t-"signal" edges according to the type of the site.
  - $t = (I)$: an edge $(v_{p-1,0}^r, v_{p,F(p)}^l)$ is added to the graph. It represents the fact that an ATG in frame $F(p)$ allows to go from intergenic to exonic in frame $F(p)$.

- $t = (T)$: an edge $(v^r_{p-1,F(p)}, v^l_{p,0})$ is added to the graph. Furthermore, the edge $(v^r_{p-1,F(p)}, v^l_{p,F(p)})$ is deleted. This represents the fact that an in frame Stop codon stops translation.
- $t = (D)$: an edge $(v^r_{p-1,i}, v^l_{p,((F(p)-i+3) \bmod 3)+4})$ is added for all $i, 1 \leq i \leq 3$. This represents the fact that splicing an exon in frame i using a donor site in frame $F(p)$ leads to an intron in phase $1 + ((F(p) - i + 3) \bmod 3)$.
- $t = (A)$: an edge $(v^r_{p-1,i+3}, v^l_{p,((F(p)+i+1) \bmod 3)+1})$ is added for all $i, 1 \leq i \leq 3$. This represents the fact that after an intron in phase i using an acceptor site in frame $F(p)$ leads to an exon in frame $1 + ((F(p) + i + 1) \bmod 3)$.
- similar modifications are done for sites occurring on the reverse complement strand.

Considering weights, we just assume that all edges $e \in E$ are weighted with a weight $w(e) \in [0, +\infty]$. As we mentioned before, in order to identify a shortest path in this graph, the simplest algorithm is the so-called Bellman algorithm (see [1] or [4], Sect. 25.4).This algorithm is the most simple instance of dynamic programming. It exploits the fact that any sub-path of an optimal path must also be optimal. This algorithm associates to each vertex $u \in V$ a variable $SP[u]$ that will *in fine* contain the length of a shortest path from s to u and a variable $\pi[u]$ that will contain the vertex that must be used to reach u in a shortest path. Initially, only $SP[s]$ is known and equal to 0. The length of a shortest path from s to v is simply the minimum over all u immediately preceding v of the sum of the length of the shortest path from s to u and the weight $w((u, v))$. The parent $\pi[v]$ is then fixed to the the vertex u that minimises this sum. The graph being acyclic, any topological ordering can be used to apply this simple relaxation procedure iteratively and finally compute $SP[t]$. This computation can be made in a time and space linear in the size of the graph and therefore in the length of the sequence (linear space is needed for the $\pi[v]$ data structure, needed to recover the shortest path at the end).

The minimum length constraints we want to take into account can be formalised as follows. A minimum length constraint is defined by a length ℓ and a track number $k, -6 \leq k \leq 6$. A prediction is a path from s to t in the prediction graph. For any path in the prediction graph, we say it violates a constraint $\langle \ell, k \rangle$ if it contains a sub-path $\langle u_0, u_1, \ldots, u_{m-1}, u_m \rangle$ such that:

- $u_0 \neq s$ and has a track different from k,
- $u_m \neq t$ and has a track different from k,
- for all vertices $u_i, 1 \leq i \leq m - 1$, u_i's track is k
- and finally $m - 1 < 2k$

Such a sub-path is said to be a violating sub-path, the edge (u_0, u_1) being its opening edge and (u_{m-1}, u_m) being its closing edge. A path is said to be feasible if it does not violate any constraint.

The algorithm exploits the following property:

Property 1. Let P be an optimal feasible path from s to a given vertex v and (u, v) be the last edge of P.

1. if v and u are on the same track, then the sub-path that connects s to u is also an optimal feasible path from s to u.
2. else, the sub-path of P that connects s to u is either entirely composed of vertices from track k (except s) or it must finish by a path $\langle u_1, \ldots, u_{m-1}, u \rangle$ that remains on track k for at least $m = 2k$ vertices. In both cases, the sub-path from s to u_1 must be a feasible optimal path from s to u_1.

Proof. In the first case, the sub-path from s to u must be feasible optimal or else we could use a shorter feasible sub-path from s to u as a short-cut and improve P without changing feasibility. We get the usual Bellman's property.

In the second case, this is no more true because an optimal feasible path from s to u may finish by a sub-path such that the addition of edge (u, v) will be the closing edge of a sub-path that violates a constraint $\langle \ell, k \rangle$. Since we want to consider only feasible paths, we can conclude that the the sub-path of P that connects s to u is either entirely composed of vertices from track k (except s) which means that it does not contain an opening edge or that it must finish by a path $\langle u_1, \ldots, u_{m-1}, u \rangle$ that remains on track k for at least $m = 2k$ vertices. In both cases, the sub-path from s to u_1 must be a feasible optimal path from s to u_1 or else we could shortcut. We get an adapted version of Bellman's property.

To exploit this property, EuGène associates two variables to each vertex u in the graph: $SP[u]$ contains the length of the shortest feasible path that goes from s to u, $\pi[u]$ contains a reference to a vertex that must be used to reach u using a shortest feasible path. Initially, only $SP[s]$ is known and equal to 0. Consider a feasible shortest path from s to a given vertex v: it must reach v using an edge (u, v) that connect one of the vertices that immediately precede v in the graph to v.

- if u and v are on the same track, we know by property 1.1 that the shortest feasible path that reach v through u, noted $SP_u[v]$ has length $SP[u] + w((u, v))$. In this case, we say that u is the u-parent of v.
- else, the second case of property 1.1 applies. Let k be the track of u and $\langle \ell, k \rangle$ the constraint that applies to track k (if no constraint exists, one can use the trivial $\langle 0, k \rangle$ constraint). The shortest feasible path that reach v through u can simply be obtained by going back on track k for $2k$ vertices, reaching a vertex w on track k (if vertex v is too close to s, we will reach the vertex $w = v_{0,k}^r$ and stop there). Let $d_k[w, u]$ be the length of the path that goes from w to u staying on track k. Since k is bounded, both w and $d_k[w, u]$ can be computed in constant time. The length of a feasible shortest path that reach v from u is then $SP_u[v] = SP[w] + d_k[w, u] + w((u, v))$, which again can be computed in constant time. In this case, we say that w is the u-parent of v.

Consider all the vertices u immediately preceding v. The number of such vertices is bounded by the number of tracks which is fixed. We can compute all $SP_u[v]$ in constant time. We have $SP[v] = \min_u(SP_u[v])$. Let $u^* = arg\min_u(SP_u[v])$. We then set $\pi[v]$ to the u^*-parent of v.

This elementary constant time procedure can be applied iteratively from vertex s to vertex t using any topological ordering of the vertices. According to property 1.1, after processing $SP[t]$ contains the length of the shortest feasible path from s to t and this path can be recovered by following the $\pi[u]$ variables backward from t. Since $|V| = O(n)$, the algorithm is time linear in the length of the sequence. The only data-structures used being the $SP[u]$ and $\pi[u]$, it is also space linear.

1.2   Scoring

The algorithm above applies to any weighting of the graph. It is now time to describe how edges are weighted. The first version of our prototype, called EuGÈNE I, integrates the following sources of information:

- output of five interpolated Markov models (IMM, [17]) for respectively frame 1, 2, 3 exons, introns and intergenic sequences. Given the sequence, the IMMs allow to compute the probability $P_k(N_i)$ that the nucleotide N_i at position i appears on track k. The corresponding edge is weighted by $-\log(P_t(N_i))$ (see Fig. 1).

  In practice, these models have been estimated on the AraClean release 1.1 dataset [7] using maximum likelihood estimation for all orders from 0 to 8. For each conditional probability distribution, a linear combination of the distributions obtained for orders 0 to 8 is used according to the amount of information available, as in Glimmer [17].

  A difference with the IMM approach used in Glimmer is that the interpolated Markov models used to estimate the probabilities $P_t(N_i)$ are used in such a way that the graph G that represents a sequence and the graph $\bar{G}$ that represents its reverse complementary are equivalent up to a half-turn of the graph. This guarantees that any sequence will be analysed exactly as its reverse complementary which seems desirable. To do this, the Markov models on the forward strand are estimated supposing that the Markov property holds in one direction. The Markov model on the reverse strand assume that the Markov property holds in the other direction. The same matrix can therefore be used on both forward and reverse strand (after a reverse complementation) which reduces the number of coding models from 6 to 3.

- output of existing signal prediction software. These software typically output a so-called "confidence" $0 \le c_i \le 1$ on the fact that a signal occurring at position i is used (*i.e.*, the corresponding switch used). This confidence cannot decently be interpreted directly as a probability. We make the assumption that the switch's weights have the parametric form $-\log(a.c_i^b)$ where the constants a and b have to be estimated for each source of evidence (see Fig. 1).

  The a, b parameters are estimated once the IMM parameters have been estimated. The estimation is done by maximising the percentage of correct predictions by EuGÈNE on the same learning set (Araclean 1.1). For fixed values of the parameters to optimise, the measure of the percentage of correct predictions is performed by applying EuGène on the Araclean dataset. Since EuGène is linear time, this is relatively efficient. The parameter estimation is then done using a very simple genetic algorithm whose results are refined by locally sampling the parameter space (which has a reasonable number of dimensions). The whole process is brutal and relatively inefficient but is performed only once. More work is needed to see how this estimation process could be made computationally less brutal.

A second version, called EuGène II can use, in conjunction with these basic pieces of information, results from cDNA/EST and protein databases search:

– cDNA/EST splices alignments are used to modify the graph as follows: matches (resp. gaps) delete/penalise intronic (resp. exonic) tracks in the graph. This forbids/penalises paths that would be incompatible with the cDNA data. In practice, cDNA/EST hits are found using Sim4 [5].
– similarly, EuGène II can exploit protein matches (found using BlastX). However, one cannot be confident enough in such information to directly use matches/gaps as constraints and we therefore simply modify the $P_t(N_i)$ using a pseudo-count scheme described below.

In practice, the structure and weights of the graph can be directly modified by the user using a very simple language that allows to include information about starts, splice sites, exonic, intronic, intergenic tracks on a per nucleotide basis. The language allows to delete or create "signal" edges in the graph and to modify the weight of any edge (either signal or content edges). To modify weights, we simply rely on a probabilistically inspired pseudo-count scheme. Initially, all edges are not only weighted by a positive weight w but also by a count c. If the user wants to modify the weight of an edge, s/he must mention a weight $w' \in [0, +\infty]$ and a count c'. The corresponding edge will then be weighted $-\log(\frac{exp(-w).c+exp(-w').c'}{c+c'})$ with count $c + c'$.

Returning to protein databases, EuGène can accept similarity information from different databases. Each similarity information taken into account modifies the weights following the previous pseudo-count scheme. If several databases are used, a specific "count" is associated with each database. These counts are optimised as the a, b parameters for signal sensors are. Using SPTR, PIR and TrEMBL as 3 different databases, we observed that the estimated count for SPTR is stronger than for the estimated count of PIR. A more frightening result is that the estimated count for TrEMBL is close to 0.

2   EuGène in Action

In this section, we show how EuGène works in practice by applying it to the contig 38 from the Araset dataset [11] which contains two genes with respectively 3 and 13 exons. We first collect information about splice sites and ATG by submitting the sequence to NetGene2 [19], SplicePredictor [2] and NetStart [12]

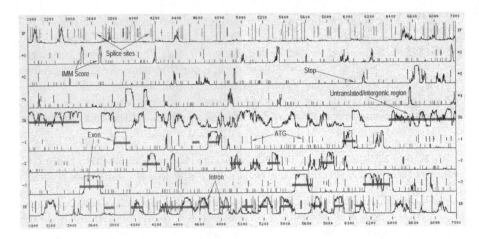

Fig. 2. EuGène I applied to contig 38 of Araset

using a dedicated Perl script. This combination of information sources is, at this time, our best "similarity-free" ingredient selection for *A. thaliana*. The script Perl automatically builds files containing positions and strengths of "switches" in the graph. Additional information may be provided by the user. For example, a line stating "`start r4 vrai 3.7e-02 nocheck`" states that there is a reverse start at position 4 and the weight of the corresponding edge should be $-\log(0.037)$. The "`nocheck`'' indicates that the user does not want EuGène to report at the end if this `ATG` has been effectively used in the prediction.

We then start EuGène and ask for a graphical zoom from nucleotide 3001 to 7000 (region of the second gene according to Araset's annotations). On this sequence, EuGène I locates correctly all exons/introns border of the 2 genes. EuGène outputs images in PNG or GIF format which can directly be used on Web pages. The X-axis is the sequence. The Y-axis represents successive "tracks": reverse introns, frame -3, -2, -1 exons, intergenic sequences (this includes UTR), frame 1, 2, 3 exons and forward introns.

On each track, the black curve represents the output of the IMM models smoothed over a window of 100 nucleotides and normalised. The large horizontal blocks represent EuGène's prediction. On the exonic tracks alone, small vertical bars represent potential stops and potential starts (`ATG`), the height of the bar being representative of the quality of the `ATG` according to the available evidence. On the intronic tracks, the vertical bars represent donors/acceptors. Again, the height of the bar is representative of the quality of the splice site according to EuGène's sources.

EuGène I is not always as successful, eg. on the sequence of a Asparagine-tRNA ligase, where EuGène I chooses one wrong splice site for one exon border. However, enough cDNA data exists so that EuGène II is able to unambiguously locate all exons/introns borders. The initial spliced alignment between the genomic sequence and the cDNA is done using `sim4` [5] and then used by EuGène.

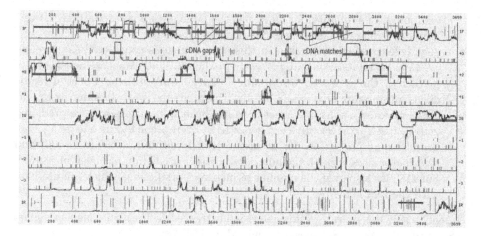

Fig. 3. EuGène II applied to a Asparagine-tRNA ligase with cDNA data

On the graphic below, this additional information is provided: the intronic tracks show blocks that represent cDNA matches interconnected by thin lines that represent gaps (connecting splice sites). In this case, no protein database information is used. When BlastX hits are fed into EuGène, they are visualised similarly on exonic tracks.

3 Evaluation

The parameters used in EuGène have been estimated on Araclean [7]. Araclean contains 144 genes with very few context around the genes. This lack of context makes it difficult for EuGène to decide whether an intergenic or intronic mode should be used on each sequence borders. Anyway, EuGène I correctly identifies all the exons (including ATG and Stop) of 76% of the sequences of Araclean. Taking into account the lack of context, if we inform EuGène I that all predictions must start and end in intergenic mode (*i.e.*, that no partial gene is allowed) then the percentage of perfectly recognised genes gets up to more than to 82%, without using any cDNA/EST/protein homology information.

This test is not very satisfactory. To get a better idea of EuGène performances, we have used another recent dataset. The "Araset" dataset has originally been designed and used to assess several existing gene or signal finding software in [11]. The clear winner of this evaluation is GeneMark.hmm [9]. We therefore compared EuGène to GeneMark.hmm on this dataset, which does not share any sequence with the Araclean dataset used for EuGène parameter estimation. The evaluation performed in [11] includes different performance measures, among which we will consider:

- sites prediction quality: for splice sites, the number of predicted, correct, over-predicted and missing sites has been measured.

- exon prediction quality: for all exons, the number of predicted, correct, overlapping, wrong and missing exons has been measured.
- gene level prediction quality: in this case, a gene is considered as correctly predicted if all its exons are correctly predicted. Il all exons are not correctly predicted, the gene can be missing (all exons are missing), partial (some exons are missing or partial), wrong (the predicted gene does not exist – as far as we know), split (a correct gene has been split in two or more predicted genes), fused (correct adjacent genes have been predicted as one gene).
- gene borders prediction: for predicted translation initiation sites (AUG) and terminator sites (stops), the number of correct prediction has been measured. For AUG, we also give the number of incorrect predictions which are in phase with the real AUG site.

In each case, the overall performance is abstracted in the two usual *specificity* (S_p) and *sensitivity* (S_n) measures introduced in [18].

S_n = number of correctly predicted items / number of actual items
S_p = number of correctly predicted items / number of predicted items

For a more detailed definition of the measures, we invite the reader to refer to [11,18]. Note also that the biological truth, considered here as unambiguously provided by Araset's annotations, is not so simple: alternative splicing or translation initiation exists and make these numbers less reliable than one may think *a priori*.

Table 1 presents the results[3] obtained for splice sites. EuGÈNE I is less sensitive but more specific than GeneMark.HMM. Table 2 gives similar numbers for translation initiation (AUG) and terminator (stop codons) sites. Here, the results are clearly in favour of EuGÈNE.

For the gene model level we also evaluated EuGÈNE II. We used SPTR as the protein database. We also have a cDNA/EST database built using EMBL, and cleaned from documented partially or alternatively spliced cDNA. However, this database is built entirely automatically and is known to contain data inconsistent with Araset annotations for some of the genes of Araset (these inconsistencies are the results of contamination with genomic DNA and undocumented alternatively or partially spliced cDNA). Although using this cDNA/EST information alone improves the performances of EuGÈNE, the version of EuGÈNE II tested uses only protein informations from SPTR because of its reliability. A clean cDNA/EST database should hopefully improve EuGÈNE II's performances.

Although an in-depth analysis is still needed to be more conclusive, it appears from the measures above that one of the strength of EuGÈNE lies in its ability to correctly identify gene's borders: compared to GeneMark.HMM, EuGÈNE merges or split very few genes and identifies most translation initiation sites or terminator sites correctly. We think that this improvement is caused by two original elements of EuGÈNE: one is the use of a dedicated intergenic Markov model,

---

[3] Reference [11] gives a number of introns (and therefore acceptors and donors) of 860. However, parsing the reference file distributed with the instances, we only found 859 introns.

Table 1. Donor and acceptor sites prediction (859 sites each)

Program	predicted	correct	overpredicted	missing	S_n	S_p	predicted	correct	overpredicted	missing	S_n	S_p
GeneMark	981	797	184	63	0.93	0.81	914	772	142	88	0.90	0.84
EuGène I	798	742	56	117	0.86	0.93	801	733	68	126	0.85	0.92

Table 2. Translation initiation and terminator sites prediction (168 sites each)

Program	predicted	correct	in phase	S_n	S_p	predicted	correct	S_n	S_p
GeneMark	196	129	31	0.77	0.66	173	136	0.81	0.79
EuGène I	190	143	18	0.85	0.75	193	155	0.92	0.80

Table 3. Exons prediction

Program	predicted	correct	overlap	wrong	missing	S_n	S_p	ratio wrong	split	fused
GeneMark	1104	845	172	87	26	0.82	0.77	0.08	10	4
EuGène I	991	837	102	57	89	0.82	0.84	0.06	1	5

Table 4. Gene level measures

Program	actual	predicted	correct	missing	partial	wrong	split	fused	S_n	S_p
GeneMark	168	208	67	1	100	27	18	12	40%	32 %
EuGène I	168	194	104	1	63	19	8	2	62%	54 %
EuGène II	168	198	125	1	42	22	9	0	74%	63 %

which is not a background 0th order model and is different from the intronic model; the other lies in the fact that, except for the IMM, its parameters have been estimated by maximum of "successful recognition" rather than maximum likelihood. This probably tends to compensate for the weaknesses of the model. Initially, we thought that the good quality of the splice sites predictors used by EuGène (NetGene2 and Splice Predictor) was a major reason for its good performances. Looking to the splice sites prediction ability of EuGène, this is less obvious now although EuGène clearly inherits the excellent specificity of NetGene2.

This report is still quite preliminary and we expect to enhance EuGÈNE's effectiveness in a near future (and apply it to other organisms). Actually, compared to other gene finding algorithms, EuGÈNE is relatively simple: it uses a single Markov model set independently of GC% (although this is probably more important for the human genome than for *Arabidopsis thaliana*), does not take into account signals such as polyA or promoters or does not model UTR. This should leave room for improvements. Another important direction lies in the use of protein and cDNA/EST hits. We think that the integration, at the lowest level, of spliced alignment algorithms in the spirit of [5,20] could help both to better take into account similarity information and also to filter out most inconsistent data. Other sources of similarity such as conservation of exons between (orphan) genes of the same family should also be integrated (M-F. Sagot, personal communication). We are currently working on these points.

Binaries for EuGÈNE can be requested from Thomas.Schiex@toulouse.inra.fr. The parasitic behaviour of EuGÈNE is a weakness here: to have a complete running version of EuGÈNE, one must first install NetStart, SplicePredictor and NetGene2 which itself depend on SAM. However, for light users, a Perl script that submit one sequence to the web interfaces of each ofl these softwares and that collects the results in the adequate file format is made available. Because EuGÈNE parameter estimation has been done using precise versions of these softwares, the best idea is, *a priori*, to install the version used for estimation. After a local installation of the above tools using NetGene2 rel. 2.42 and SAM rel. 3.2, we got the bad surprise that NetGene2 did not gave the same output as the Web version which we used up to now. The nice surprise is that without any a, b parameter re-estimation, the gene-level sensibility and specificity have *increased* to 64.2% and 57.7% respectively. This shows both the good robustness of EuGÈNE and a positive side-effect of its parasitic behaviour: any improvement in the hosts may yield an improvement in EuGÈNE itself.

Acknowledgements. This work has been supported by a grant from the " Programme prioritaire génome & fonction " of the *Institut national de la recherche agronomique* (INRA). We would also like to thank N. Pavie and C. Mathé (INRA, Gand) for their help with Araset, S. Rombauts (INRA, Gand), B. Lescure (INRA, Toulouse), H. Chiapello and I. Small (INRA, Versailles) for sharing their experience de l'arabette, P. Dehais and S. Aubourg (INRA, Gand) for their feedback and their help in making EuGÈNE more user friendly.

References

[1] R. Bellman, *Dynamic Programming*, Princeton Univ. Press, Princeton, New Jersey, (1957).

[2] V. Brendel and J. Kleffe, (1998), *Prediction of locally optimal splice sites in plant pre-mRNA with application to gene identification in Arabidopsis thaliana genomic DNA*, Nucleic Acids Res., 26, pp. 4749–4757.

[3] C. Burge and S. Karlin, Apr 1997, *Prediction of complete gene structures in human genomic dna.*, J Mol Biol, 268, pp. 78–94.

[4] T. H. Cormen, C. E. Leiserson, and R. L. Rivest, *Introduction to algorithms*, MIT Press, (1990). ISBN : 0-262-03141-8.

[5] L. Florea, G. Hartzell, Z. Zhang, G. Rubin, and W. Miller, Sept. 1998, *A computer program for aligning a cdna sequence with a genomic dna sequence*, Genome Res, 8, pp. 967–974.

[6] X. Huang, M. Adams, H. Zhou, and A. Kerlavage, Nov 1997, *A tool for analyzing and annotating genomic sequences.*, Genomics, 46, pp. 37–45.

[7] P. Korning, S. Hebsgaard, P. Rouze, and S. Brunak, (1996), *Cleaning the genbank arabidopsis thaliana data set*, Nucleic Acids Res., 24, pp. 316–320.

[8] D. Kulp, D. Haussler, M. Reese, and F. Eeckman, (1997), *Integrating database homology in a probabilistic gene structure model.*, in Pacific Symp. Biocomputing, pp. 232–44.

[9] A. V. Lukashin and M. Borodovsky, (1998), *GeneMark.hmm: new solutions for gene finding*, Nucleic Acids Res., 26, pp. 1107–1115.

[10] K. Murakami and T. Takagi, (1998), *Gene recognition by combination of several gene-finding programs*, BioInformatics, 14, pp. 665–675.

[11] N. Pavy, S. Rombauts, P. Déhais, C. Mathé, D. Ramana, P. Leroy, and P. Rouzé, Nov. 1999, *Evaluation of gene prediction software using a genomic data set: application to arabidopsis thaliana sequences.*, Bioinformatics, 15, pp. 887–99. Also appeared in the Proc. of 2^d Georgia Tech conference on BioInformatics.

[12] A. Pedersen and H. Nielsen, (1997), *Neural network prediction of translation initiation sites in eukaryotes: prespectives for EST and genome analysis*, in Proc. of ISMB'97, AAAI Press, pp. 226–233.

[13] G. R, Winter 1998, *Assembling genes from predicted exons in linear time with dynamic programming.*, Journal of Computational Biology, 5, pp. 681–702.

[14] L. Rabiner, (1989), *A tutorial on hidden markov models and selected application in speech recognition*, Proc. IEEE, 77, pp. 257–286.

[15] L. R. Rabiner, (1989), *A tutorial on hidden markov models and selected applications in speech recognition*, Proc. of the IEEE, 77, pp. 257–286.

[16] I. Rogozin, L. Milanesi, and N. Kolchanov, Jun 1996, *Gene structure prediction using information on homologous protein sequence.*, Comput. Appl. Biosci., 12, pp. 161–70.

[17] S. L. Salzberg, A. L. Delcher, S. Kasif, and O. White, (1998), *Microbial gene identification using interpolated Markov models*, Nucleic Acids Res., 26, pp. 544–548.

[18] E. Snyder and G. Stormo, *DNA and protein sequence analysis: a practical approach*, IRL Press, Oxford, (1995), ch. Identifying genes in genomic DNA sequences, pp. 209–224.

[19] N. Tolstrup et al., (1997), *A branch-point consensus from Arabidopsis found by non circular analysis allows for better prediction of acceptor sites*, Nucleic Acids Res., 25, pp. 3159–3163.

[20] J. Usuka., W. Zhu., and V. Brendel., (2000), *Optimal spliced alignment of homologous cDNA to a genomic DNA template*, Bioinformatics, 16, pp. 203–211.

[21] T. D. Wu, (1996), *A segment-based dynamic programming algorithm for predicting gene structure*, Journal of Computational Biology, 3, pp. 375–394.

Tree Reconstruction via a Closure Operation on Partial Splits*

Charles Semple and Mike Steel

Biomathematics Research Centre
Department of Mathematics and Statistics
University of Canterbury
Private Bag 4800
Christchurch, New Zealand
{c.semple,m.steel}@math.canterbury.ac.nz

Abstract. A fundamental problem in biological classification is the reconstruction of phylogenetic trees for a set X of species from a collection of either subtrees or qualitative characters. This task is equivalent to tree reconstruction from a set of partial X-splits (bipartitions of subsets of X). In this paper, we define and analyse a "closure" operation for partial X-splits that was informally proposed by Meacham [5]. In particular, we establish a sufficient condition for such an operation to reconstruct a tree when there is essentially only one tree that displays the partial X-splits. This result exploits a recent combinatorial result from [2].

1   Introduction

Trees that have some vertices labelled by elements from a finite set X are often used to represent evolutionary relationships, particularly in biology. Two closely related problems are

(i) determining how to combine such trees that classify overlapping subsets of X into a parent tree that displays each of the input trees, and

(ii) determining how to reconstruct a parent tree from (partial) qualitative characters (equivalently, partitions of X or subsets of X) so that each character could have evolved on the parent tree without any reverse or convergent transitions (this is equivalent to each tree displaying the partition associated with each character).

For either problem, a parent tree may not exist, and even deciding this turns out to be an NP-complete problem [3,6]. However, in certain cases, various efficient "rules" for extending sets of trees or sets of characters can either determine that such a tree does not exist, or reconstruct the tree when there is essentially only one possible parent tree. Conditions for such an approach to succeed using extension rules on sets of trees were recently established in [1], using a combinatorial result from [2]. In this paper we take a related, but different, approach by

---

* This work was supported by the New Zealand Marsden Fund (UOC-MIS-003).

O. Gascuel, M.-F. Sagot (Eds.): JOBIM 2000, LNCS 2066, pp. 126–134, 2001.

considering a rule for extending sets of partial binary characters (called partial X–splits below) that was proposed informally by Meacham [5]. We formalise an iterative construction using this rule, and show that it always leads to the same set of partial X–splits, regardless of the possible choices by which the rule can be applied. Then, using the main combinatorial result from [2], we provide sufficient conditions for this construction to successfully recover a parent tree or determine that no such tree exists. Note that although the input to our tree reconstruction problem consists of partial X–splits, it could easily be modified to input partitions of subsets of X (in the case of problem (ii)) or trees classifying overlapping subsets of X (in the case of problem (i)) since all these problems are essentially equivalent [6].

2   Preliminaries

Throughout this paper, X denotes a finite set. We begin with some definitions.
Partial splits. A *partial split* of X, or more briefly a *partial* X*–split*, is a partition of a subset of X into two disjoint non-empty subsets. If these two subsets are A and B, we denote the partial split by $A|B$. Note that no distinction is made between $A|B$ and $B|A$. If $A \cup B = X$ we say that $A|B$ is a *(full)* X*–split*. We write $aa'|bb'$ to denote the partial split $A|B$ if $A = \{a, a'\}$ and $B = \{b, b'\}$, and we call this a *quartet* X*–split*. We say that the partial split $A'|B'$ *extends* the partial split $A|B$ precisely if either $A \subseteq A'$ and $B \subseteq B'$ or $A \subseteq B'$ and $B \subseteq A'$. A partial X–split $A|B$ is *trivial* if $\min\{|A|, |B|\} = 1$.
X–trees. Let T be a tree with vertex set V and edge set E, and suppose we have a map $\phi : X \to V$ with the property that, for all $v \in V$ with degree at most two, $v \in \phi(X)$. Then the ordered pair $(T; \phi)$, which we frequently denote by $\mathcal{T}$, is called an X*–tree*. Two X–trees $(T_1; \phi_1)$ and $(T_2; \phi_2)$, where $T_1 = (V_1, E_1)$ and $T_2 = (V_2, E_2)$, are regarded as equivalent if there exists a bijection $\psi : V_1 \to V_2$ which induces a bijection between E_1 and E_2 and satisfies $\phi_2 = \psi \circ \phi_1$, in which case, ψ is unique.

Let $\mathcal{T} = (T; \phi)$ be an X–tree and let e be an edge of T. Then *corresponding* to e is the X–split $\phi^{-1}(V_1)|\phi^{-1}(V_2)$, where V_1 and V_2 denote the vertex sets of the two components obtained from T by deleting e. For an X–tree $\mathcal{T}$, let $\Sigma(\mathcal{T})$ (resp. $\Sigma^*(\mathcal{T})$) denote the collection of non-trivial X–splits (resp. all X–splits) corresponding to the edges of T.
Compatibility. Let $A|B$ be a partial X–split. An X–tree $\mathcal{T} = (T; \phi)$ *displays* $A|B$ if there is an edge e of $T = (V, E)$ such that, in $(V, E - \{e\})$, the sets $\phi(A)$ and $\phi(B)$ are subsets of the vertex sets of different components. For example, the X-tree shown in Fig. 1, where $X = \{1, 2, \dots, 7\}$, displays each of the partial X–splits in $\{\{1, 2\}|\{3, 4\}, \{2, 3\}|\{4, 7\}, \{1, 7\}|\{4, 5\}, \{2, 5\}|\{6, 7\}\}$. A collection Σ of partial X–splits is said to be *compatible* if there exists an X–tree that displays every X–split in Σ. This is equivalent to requiring that every non-trivial split in Σ is extended by a split in $\Sigma(\mathcal{T})$.

The following result is well known, and follows immediately from results in [4].

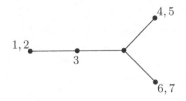

Fig. 1. An X–tree displaying $\{1,2\}|\{3,4\}, \{2,3\}|\{4,7\}, \{1,7\}|\{4,5\}$, and $\{2,5\}|\{6,7\}\}$.

Lemma 1. *Let $A_1|B_1$ and $A_2|B_2$ be partial X–splits. The following statements are equivalent:*

(i) $A_1|B_1$ *and* $A_2|B_2$ *are compatible.*
(ii) *At least one of the sets $A_1 \cap A_2$, $A_1 \cap B_2$, $B_1 \cap A_2$, and $B_1 \cap B_2$ is empty.*

A set Σ of partial X–splits is said to be *pairwise compatible* if each pair of splits in Σ is compatible. This condition is not sufficient for Σ to be compatible. For example, for $X = \{a, b, c, d, e\}$, the set $\{ab|cd, ab|ce, ad|be\}$ of partial X–splits of X is pairwise compatible, but not compatible. However, if Σ consists of full X-splits, then Σ is compatible precisely if Σ is pairwise compatible, in which case there is a unique X–tree $\mathcal{T}$ such that $\Sigma^*(\mathcal{T}) = \Sigma$ (see [4]).

Irreducible sets of partial X–splits. Let Σ be a set of partial splits of X. A partial split $A|B \in \Sigma$ is *redundant* if there exists a different partial split in Σ that extends $A|B$. If Σ has no redundant splits, then Σ is said to be *irreducible*. Let Σ_1 and Σ_2 be two irreducible sets of partial splits of X. We write $\Sigma_1 \preceq \Sigma_2$ if, for each $A_1|B_1 \in \Sigma_1$, there is an element $A_2|B_2$ in Σ_2 that extends $A_1|B_1$. It is not difficult to show that $\preceq$ is a partial order on the collection of irreducible sets of partial X–splits. Note that if we drop the irreducibility condition, then $\preceq$ may fail to satisfy the antisymmetric property ($a \preceq b$ and $b \preceq a$ implies $a = b$) required of a partial order. Observe that an X–tree $\mathcal{T}$ displays a set Σ of partial X–splits precisely if $\Sigma \preceq \Sigma^*(\mathcal{T})$.

We will let $\mathcal{P}(X)$ denote the collection of all sets Σ of partial X–splits that are both pairwise compatible and irreducible. Let us adjoin to $\mathcal{P}(X)$ a new element ω, and let $\mathcal{P}_\omega(X) = \mathcal{P}(X) \cup \{\omega\}$. If we extend the definition of $\preceq$ by setting $\Sigma \preceq \omega$ for all $\Sigma \in \mathcal{P}(X)$ (so that ω acts as a maximal element), then $\preceq$ is a partial order on $\mathcal{P}_\omega(X)$.

3   Split Closure

In this section, we define a "split closure" of a set Σ of partial X–splits. Informally, we construct an irreducible set of partial X–splits from Σ by repeatedly applying a pairwise replacement rule along the lines suggested by Meacham [5] (the replacement rule (SC) below corresponds to "Rule 2" in [5]). We then show that any two split closures of Σ are equal.

The replacement rule we consider for an irreducible set Σ of partial X–splits is the following:

(SC) If $A_1|B_1$ and $A_2|B_2$ are elements of Σ that satisfy

$$\emptyset \notin \{A_1 \cap A_2, A_1 \cap B_2, B_1 \cap B_2\} \text{ and } B_1 \cap A_2 = \emptyset, \tag{1}$$

replace $A_1|B_1$ and $A_2|B_2$ in Σ by $(A_1 \cup A_2)|B_1$ and $A_2|(B_1 \cup B_2)$, and then remove any redundant partial splits from the newly created set.

If $A_1|B_1$ and $A_2|B_2$ in the statement of (SC) have the property that $A_2 \subseteq A_1$, $B_1 \subseteq B_2$, and (1) applies, then $B_1 \cap A_2$ is empty, and the two newly created partial splits are $(A_1 \cup A_2)|B_1$ and $A_2|(B_1 \cup B_2)$, which are identical to $A_1|B_1$ and $A_2|B_2$, respectively. We call such an application of (SC) *trivial*; in all other (*non-trivial*) applications of (SC) at least one of the newly created partial splits differs from $A_1|B_1$ or $A_2|B_2$.

We say that $\Sigma \in \mathcal{P}(X)$ is *closed* under (SC) if (SC) applies only trivially to Σ.

The motivation for (SC) is the following result due to Meacham [5].

Lemma 2. *Let Σ be a set of partial X–splits, and let Σ' be a set of partial X–splits obtained from Σ by a single application of* (SC). *Then an X–tree T displays Σ if and only if T displays Σ'.*

Let Σ be a set of irreducible partial X–splits, and suppose that we construct a sequence

$$\Sigma_0, \Sigma_1, \ldots, \Sigma_i, \Sigma_{i+1}, \ldots$$

of irreducible partial X–splits such that $\Sigma_0 = \Sigma$ and, for all $i \geq 0$, Σ_{i+1} is obtained from Σ_i by one non-trivial application of (SC) provided Σ_i is pairwise compatible. Since $\Sigma_i \preceq \Sigma_{i+1}$, for all $i \leq 0$, it follows that this sequence is strictly increasing under $\preceq$. Consequently, since the set of all X–splits is finite, this sequence must terminate with a set, Σ_n say, of irreducible partial X–splits such that either Σ_n is pairwise compatible and closed under (SC), or Σ_n is not pairwise compatible. If the latter holds, we reset Σ_n to be the element ω.

Definition. We refer to the sequence $\Sigma_0, \Sigma_1, \ldots, \Sigma_n$ as a *split closure sequence* for Σ, and the terminal value Σ_n as a *split closure* of Σ. Note that (SC) applies only trivially to Σ_n (when $\Sigma_n \neq \omega$) and Σ_n is always an upper bound, under $\preceq$, to Σ.

We next provide an explicit bound on the length of any split closure sequence.

Lemma 3. *Let Σ be a set of irreducible partial X–splits, and let $\Sigma_0, \Sigma_1, \ldots, \Sigma_n$ be a split closure sequence for Σ. Then $n \leq |\Sigma| \times |X| - \sum_{A|B \in \Sigma} |A \cup B|$.*

Proof. It is straightforward to see that we can prove the lemma by making the additional assumption that $\Sigma_n \neq \omega$. For all $i \in \{0, 1, \ldots, n-1\}$, let $\lambda_i : \Sigma_i \to \Sigma_{i+1}$ be a function that maps an element, $A'|B'$ say, of Σ_i to an element of Σ_{i+1} that extends $A'|B'$. Furthermore, for each element, $A|B$ say, of Σ and for all $i \in \{0, 1, \ldots, n-1\}$, let $A_{i+1}|B_{i+1} = \lambda_i \lambda_{i-1} \cdots \lambda_0(A|B)$.

Since, for all i, Σ_{i+1} is obtained from Σ_i by a non-trivial application of (SC) and $\Sigma_i \preceq \Sigma_{i+1}$, it follows that

$$\sum_{A|B \in \Sigma} (|A_{i+1} \cup B_{i+1}| - |A_i \cup B_i|) \geq 1,$$

for all i. Consequently,

$$\sum_{A|B \in \Sigma} (|A_n \cup B_n| - |A \cup B|) \geq n.$$

Therefore, as $|A_n \cup B_n| - |A \cup B| \leq |X| - |A \cup B|$ for each element $A|B$ in Σ,

$$n \leq |\Sigma| \times |X| - \sum_{A|B \in \Sigma} |A \cup B|$$

as required. □

It will immediately follow from Lemma 4 that the split closure of a set Σ of irreducible partial X–splits is well-defined.

Lemma 4. *Let Σ be an irreducible set of partial X–splits. Then any two split closures of Σ are equal.*

Proof. If every split closure of Σ is ω, the lemma is (trivially) true, so we may assume that there exists a split closure, $\overline{\Sigma}$ say, of Σ which is not ω. We prove the lemma by showing that every other split closure of Σ equals $\overline{\Sigma}$. To this end, let $\Sigma_0, \Sigma_1, \ldots, \Sigma_n$ be a split closure sequence for Σ, where $\Sigma_0 = \Sigma$. We first claim that, for all $i \in \{0, 1, \ldots, n\}$,

$$\Sigma_i \neq \omega \text{ and } \Sigma_i \preceq \overline{\Sigma}. \tag{2}$$

We establish (2) by induction on i. If $i = 0$, then (2) holds as there exists a split closure of Σ not equal to ω. Now suppose that (2) holds for $i = r$, where $r \in \{0, 1, \ldots, n-1\}$, and Σ_{r+1} is obtained from Σ_r by applying (SC) to the pair $A_1|B_1$ and $A_2|B_2$. Without loss of generality, we may assume that $B_1 \cap A_2 = \emptyset$. By the induction hypothesis, $\Sigma_r \preceq \overline{\Sigma}$, and so there is a pair of partial X–splits $A_1'|B_1'$ and $A_2'|B_2'$ in $\overline{\Sigma}$ such that $A_i \subseteq A_i'$ and $B_i \subseteq B_i'$, for all $i \in \{1, 2\}$. Since $\overline{\Sigma}$ is pairwise compatible, it follows that $A_1'|B_1'$ and $A_2'|B_2'$ satisfy (1). Therefore, as (SC) applies only trivially to $\overline{\Sigma}$, it follows that $A_2' \subseteq A_1'$ and $B_1' \subseteq B_2'$. Consequently, $A_1'|B_1'$ and $A_2'|B_2'$ extend $(A_1 \cup A_2)|B_1$ and $A_2|(B_1 \cup B_2)$, respectively, and so

$$\Sigma_r \cup \{(A_1 \cup A_2)|B_1, A_2|(B_1 \cup B_2)\} - \{A_1|B_1, A_2|B_2\} \preceq \overline{\Sigma}.$$

Therefore, as $\overline{\Sigma}$ is pairwise compatible, $\Sigma_{r+1} \neq \omega$ and $\Sigma_{r+1} \preceq \overline{\Sigma}$. This completes the induction step and thereby establishes (2).

Applying (2) to $i = n$, we get $\Sigma_n \neq \omega$ and $\Sigma_n \preceq \overline{\Sigma}$. By interchanging the roles of Σ_n and $\overline{\Sigma}$ in the argument of the last paragraph, we deduce that $\overline{\Sigma} \preceq \Sigma_n$, and hence $\Sigma_n = \overline{\Sigma}$. □

Definition. In view of Lemma 4, we denote the split closure of a set Σ of irreducible partial X–splits by $\mathrm{spcl}(\Sigma)$.

Note that $|\mathrm{spcl}(\Sigma)| \leq |\Sigma|$. Also, provided we set $\mathrm{spcl}(\omega) = \omega$, then spcl satisfies the three properties one expects of a closure operation on the poset $\mathcal{P}_\omega(X)$. Namely, if $a, b \in \mathcal{P}_\omega(X)$ with $a \preceq b$, then

(i) $a \preceq \mathrm{spcl}(a)$,
(ii) $\mathrm{spcl}(a) \preceq \mathrm{spcl}(b)$, and
(iii) $\mathrm{spcl}(\mathrm{spcl}(a)) = \mathrm{spcl}(a)$.

The next result follows immediately from Lemma 2, however, the converse of this corollary is not true [6].

Corollary 1. *Let Σ be a set of irreducible partial X–splits. If $\mathrm{spcl}(\Sigma) = \omega$, then Σ is incompatible.*

4   Tree Reconstruction Using Split Closure

In this section, we establish a sufficient condition for the split closure of an irreducible set of partial X–splits to recover all the non-trivial splits of an X–tree. This result, Corollary 2, depends on a combinatorial theorem from [2]. In order to apply this theorem, we need to relate partial splits and split closure to quartet splits and a dyadic closure rule that operates on quartet splits. To this end, we introduce some further definitions.

Definition. For a partial X–split $A|B$, let

$$\mathcal{Q}(A|B) = \{aa'|bb' : a, a' \in A; \; b, b' \in B; \; a \neq a'; \text{ and } b \neq b'\}$$

and, for a set Σ of partial X–splits, let

$$\mathcal{Q}(\Sigma) = \bigcup_{A|B \in \Sigma} \mathcal{Q}(A|B).$$

For an X-tree $\mathcal{T}$, we denote $\mathcal{Q}(\Sigma(\mathcal{T}))$ by $\mathcal{Q}(\mathcal{T})$.

Proposition 1 relates partial splits to quartet splits.

Proposition 1. *Let Σ be an irreducible set of non-trivial partial X–splits and let $\mathcal{T}$ be an X–tree. Then $\Sigma = \Sigma(\mathcal{T})$ if and only if the following two conditions hold:*

(i) $|\Sigma| \leq |\Sigma(\mathcal{T})|$; *and*
(ii) $\mathcal{Q}(\Sigma) = \mathcal{Q}(\mathcal{T})$.

Proof. Evidently, if $\Sigma = \Sigma(\mathcal{T})$, then (i) and (ii) hold. For the converse, we first show that $\Sigma \preceq \Sigma(\mathcal{T})$. Let $\mathcal{T} = (T; \phi)$, and let $A|B$ be an element of Σ. By (ii), $\mathcal{Q}(A|B) \subseteq \mathcal{Q}(\mathcal{T})$. Therefore, for each quartet of elements a, a', b, and b' with $a, a' \in A$ and $b, b' \in B$, the cardinality, denoted $n(a, a', b, b')$, of $\{A'|B' \in \Sigma(\mathcal{T}) : aa'|bb' \in \mathcal{Q}(A'|B')\}$ satisfies $n(a, a', b, b') \geq 1$. Now suppose that a, a', b, and b' are chosen so that $n(a, a', b, b')$ is minimised, and $A'|B'$ is an element of $\Sigma(T)$ with $aa'|bb' \in Q(A'|B')$. By considering the placement of the vertices $\phi(a)$, $\phi(a')$, $\phi(b)$, and $\phi(b')$ in T, we see that $A \subseteq A'$ and $B \subseteq B'$, thus showing that $\Sigma \preceq \Sigma(T)$.

Now let $n(A|B) = \min\{n(a, a', b, b') : a, a' \in A; b, b' \in B\}$, and let

$$\Sigma_1 = \{A|B \in \Sigma : n(A|B) = 1\}.$$

Using the fact that $\Sigma \preceq \Sigma(T)$, it is easily seen that, for each element, $A|B$ say, of Σ_1, there is a unique element of $\Sigma(T)$ that extends $A|B$. Let $\mu : \Sigma_1 \to \Sigma(T)$ denote the map that associates with each element $A|B$ of Σ_1 the unique element of $\Sigma(T)$ that extends $A|B$. We next show that μ is a bijection.

Let $C'|D'$ be an element of $\Sigma(T)$, and choose elements $c, c' \in C'$ and $d, d' \in D'$ so that $n(c, c', d, d') = 1$. Then, by (ii), there is an element $C|D$ of Σ_1 such that $cc'|dd' \in Q(C|D)$ and, moreover, $\mu(C|D) = C'|D'$. Thus the map μ is surjective and so $|\Sigma_1| \geq |\Sigma(T)|$. It now follows from (i) that $\Sigma_1 = \Sigma$, and so μ is indeed a bijection. Hence $|\Sigma| = |\Sigma(T)|$. Since $\Sigma_1 = \Sigma$, we can complete the proof by showing that, for each $A|B \in \Sigma_1$, $\mu(A|B) = A|B$.

Suppose, to the contrary, that $\mu(A|B) = A'|B'$, where $A'|B'$ extends $A|B$ but is not equal to $A|B$, for some $A|B \in \Sigma_1$. Then there is an element x in $(A' \cup B') - (A \cup B)$. Without loss of generality, we may assume $x \in A'$. Then we can choose elements $a_1 \in A'$ and $b_1, b_2 \in B'$ so that $n(x, a_1, b_1, b_2) = 1$. By (ii), there is an element $C|D$ of Σ_1 such that $xa_1|b_1b_2 \in Q(C|D)$. Since $x \notin A \cup B$, $C|D$ is not equal to $A|B$. Therefore, as μ is a bijection, $\mu(C|D) \neq A'|B'$, and so $xa_1|b_1b_2 \notin Q(\mu(C|D))$. This contradiction completes the proof of Proposition 1.
□

Following [1], the *semi-dyadic closure* of a collection Q of quartet X–splits, denoted $scl_2(Q)$, is the minimal set of quartet X–splits that contains Q and is closed under the following rule:

(SDC) If $ab|cd$ and $ac|de$ are elements of $scl_2(Q)$, then $ab|ce$, $ab|de$, and $bc|de$ are elements of $scl_2(Q)$.

The next proposition relates split closure to semi-dyadic closure.

Proposition 2. *If Σ is a set of compatible irreducible partial X–splits, then $scl_2(Q(\Sigma)) \subseteq Q(spcl(\Sigma))$.*

Proof. We can obtain $scl_2(Q(\Sigma))$ by constructing a sequence $Q_0, Q_1, \ldots, Q_m$ of collections of quartet X–splits such that $Q_0 = Q(\Sigma)$, $Q_m = scl_2(Q(\Sigma))$, and, for all $i \in \{0, 1, \ldots, m - 1\}$, $Q_{i+1} = Q_i \cup scl_2(\{q_i, q_i'\})$, where $q_i, q_i' \in Q_i$ but $scl_2(\{q_i, q_i'\}) \not\subseteq Q_i$. We prove the proposition by showing that one can construct a sequence $\Sigma_0, \Sigma_1, \ldots, \Sigma_m$ of sets of irreducible partial X–splits such that $\Sigma_0 = \Sigma$ and, for all $j \in \{0, 1, \ldots, m\}$,

$$\Sigma_j \preceq spcl(\Sigma) \text{ and } Q_j \subseteq Q(\Sigma_j). \tag{3}$$

For then, taking $j = m$, establishes the proposition.

The proof of the latter construction is by induction on j. Clearly, the result holds if $j = 0$ as $\Sigma_0 = \Sigma$. Now let r be an element of $\{0, 1, \ldots, m-1\}$, and suppose that Σ_j has been defined for all $j \le r$ and (3) holds for $j = r$. Then $q_r, q_r' \in \mathcal{Q}(\Sigma_r)$. If $\mathrm{scl}_2(\{q_r, q_r'\}) \subseteq \mathcal{Q}(\Sigma_r)$, then set $\Sigma_{r+1} = \Sigma_r$. On the other hand, suppose that $\mathrm{scl}_2(\{q_r, q_r'\}) \not\subseteq \mathcal{Q}(\Sigma_r)$. Since $\mathcal{Q}_r \subseteq \mathcal{Q}(\Sigma_r)$, there are two distinct elements $A|B$ and $A'|B'$ in Σ_r that extend q_r and q_r', respectively. As Σ is compatible, $\mathrm{spcl}(\Sigma)$ is compatible, so Σ_r is pairwise compatible. It now follows that we can apply (SC) to $A|B$ and $A'|B'$. Set Σ_{r+1} to be the resulting set of irreducible partial X–splits. In both cases, $\Sigma_{r+1} \preceq \mathrm{spcl}(\Sigma)$ and, moreover, one can easily check that $\mathrm{scl}_2(\{q_r, q_r'\}) \subseteq \mathcal{Q}(\Sigma_{r+1})$. Hence (3) holds for $j = r+1$, and so we can indeed construct such a sequence.   □

Definition. A set Σ of non-trivial partial X–splits *weakly defines* an X–tree $\mathcal{T}$ if there is a unique X–tree $\mathcal{T}'$ that displays $\Sigma \cup \{\{x\}|X - \{x\} : x \in X\}$, in which case $\Sigma^*(\mathcal{T}) = \Sigma(\mathcal{T}')$.

Let $\mathcal{T} = (T; \phi)$ be an X–tree, and let v be a vertex of T. Suppose that there is a set of non-trivial partial X–splits that weakly defines $\mathcal{T}$. Then it is easily seen that each of the following hold in T:

(i)   If v is a pendant vertex, then $|\phi^{-1}(v)| = 2$.
(ii) If v is a degree-two vertex, then $|\phi^{-1}(v)| = 1$.
(iii) If v is neither a pendant vertex nor a degree-two vertex, then v is a degree-three vertex and $\phi^{-1}(v) = \emptyset$.

Conversely, if $\mathcal{T}$ satisfies all of (i)–(iii), then $\Sigma(\mathcal{T})$ weakly defines $\mathcal{T}$. As an example, the set $\{\{1,2\}|\{3,4\}, \{2,3\}|\{4,7\}, \{1,7\}|\{4,5\}, \{2,5\}|\{6,7\}\}$ of partial X–splits weakly defines the X–tree in Fig. 1.

Two characterisations for when a minimum-sized set of quartet X–splits weakly defines an X–tree are given in [2]. Theorem 1 gives a third such characterisation. Before stating this theorem, we note that it immediately follows from [6, Proposition 6] that $|X| - 3$ is the minimum number of quartet X–splits that can weakly define an X–tree $\mathcal{T}$. Observe that $|X| - 3 = |\Sigma(\mathcal{T})|$.

Theorem 1. *Let Σ_Q be a set of $|X| - 3$ quartet X–splits, and let $\mathcal{T}$ be an X–tree. Then the following statements are equivalent:*

(i) Σ_Q *weakly defines* $\mathcal{T}$.
(ii) $\mathrm{spcl}(\Sigma_Q) = \Sigma(\mathcal{T})$.

Proof. If $\mathrm{spcl}(\Sigma_Q) = \Sigma(\mathcal{T})$, then one can easily check using Lemma 2 that Σ_Q weakly defines $\mathcal{T}$. For the converse, suppose that Σ_Q weakly defines $\mathcal{T}$. Then, by [2, Theorem 3.11] (also see [1]), $\mathrm{scl}_2(\Sigma_Q) = \mathcal{Q}(\mathcal{T})$. Since Σ_Q is compatible and irreducible, we can apply Proposition 2 to Σ_Q and get $\mathrm{scl}_2(\mathcal{Q}(\Sigma_Q)) \subseteq \mathcal{Q}(\mathrm{spcl}(\Sigma_Q))$. As $\mathcal{Q}(\Sigma_Q) = \Sigma_Q$, it follows that $\mathcal{Q}(\mathcal{T}) \subseteq \mathcal{Q}(\mathrm{spcl}(\Sigma_Q))$. Now $\mathrm{spcl}(\Sigma_Q)$ is compatible, so $\mathcal{Q}(\mathcal{T}) = \mathcal{Q}(\mathrm{spcl}(\Sigma_Q))$. Moreover, $|\mathrm{spcl}(\Sigma_Q)| \le |\Sigma_Q| = |\Sigma(\mathcal{T})|$, and so, by Proposition 1, $\mathrm{spcl}(\Sigma_Q) = \Sigma(\mathcal{T})$ as required.   □

An immediate consequence of Theorem 1 is Corollary 2.

Corollary 2. *Let Σ_Q be a set of quartet X-splits, and suppose that there exists a subset of Σ_Q of size $|X|-3$ that weakly defines an X-tree $\mathcal{T}$. If Σ_Q is compatible, then $\mathrm{spcl}(\Sigma_Q) = \Sigma(\mathcal{T})$; otherwise $\mathrm{spcl}(\Sigma_Q) = \omega$.*

Suppose that Σ_Q and $\mathcal{T}$ satisfy the assumptions of their namesake in the statement of Corollary 2. The potential utility of Corollary 2 lies in the fact that $\mathcal{T}$ can be reconstructed from $\Sigma(\mathcal{T})$ and, in turn, $\Sigma(\mathcal{T}) = \mathrm{spcl}(\Sigma_Q)$ can be reconstructed from Σ_Q; moreover, both tasks can be carried out in polynomial time. Thus we obtain an alternative polynomial-time algorithm for the special case of this tree reconstruction problem to that described in [1]. Furthermore, if $|\Sigma_Q| = O(n)$, then, by Lemma 3, every split closure sequence for Σ_Q has length at most $O(n^2)$, and so the algorithm described here should be reasonably fast.

Acknowledgements. We thank the anonymous referee for their helpful comments.

References

[1] Böcker, S., Bryant, D., Dress, A. W. M., and Steel, M. A.: Algorithmic aspects of tree amalgamation. To appear in *Journal of Algorithms*.

[2] Böcker, S., Dress, A. W. M., and Steel, M.: Patching up X-trees. *Annals of Combinatorics* **3** (1999) 1-12.

[3] Bodlaender, H. L., Fellows, M. R., and Warnow, T. J.: Two strikes against perfect phylogeny. In *Proceedings of the International Colloquium on Automata, Languages and Programming*, Vol. 623 of Lecture Notes in Computer Science, Springer-Verlag, Berlin (1993) 273–283.

[4] Buneman, P.: The recovery of trees from measures of dissimilarity. In F. R. Hodson, D. G. Kendall, and P. Tautu (eds.): *Mathematics in the Archaeological and Historical Sciences*, Edinburgh University Press (1971) 387–395.

[5] Meacham, C. A.: Theoretical and computational considerations of the compatibility of qualitative taxonomic characters. In J. Felsenstein (ed.): *Numerical Taxonomy*, NATO ASI Series Vol. G1, Springer-Verlag (1983) 304–314.

[6] Steel, M.: The complexity of reconstructing trees from qualitative characters and subtrees. *Journal of Classification* **9(1)** (1992) 91–116.

InterDB, a Prediction-Oriented Protein Interaction Database for *C. elegans*

Nicolas Thierry-Mieg and Laurent Trilling

Laboratoire LSR-IMAG,
38402, Saint-Martin-d'Hères cedex, France
{Nicolas.Thierry-Mieg,Laurent.Trilling}@imag.fr

Abstract. Protein-protein interactions are critical to many biological processes, extending from the formation of cellular macromolecular structures and enzymatic complexes to the regulation of signal transduction pathways. With the availability of complete genome sequences, several groups have begun large-scale identification and characterization of such interactions, relying mostly on high-throughput two-hybrid systems. We collaborate with one such group, led by Marc Vidal, whose aim is the construction of a protein-protein interaction map for *C. elegans*. In this paper we first describe WISTdb, a database designed to store the interaction data generated in Marc Vidal's laboratory. We then describe InterDB, a multi-organism prediction-oriented database of protein-protein interactions. We finally discuss our current approaches, based on inductive logic programming and on a data mining technique, for extracting predictive rules from the collected data.

1 The Biological Problem: Protein-Protein Interactions

Protein-protein interactions are critical to many biological processes, extending from the formation of cellular macromolecular structures and enzymatic complexes to the regulation of signal transduction pathways.

With the availability of complete genome sequences, several groups have begun large-scale identification and characterization of such interactions [11], [22], [25]. These groups rely mostly on high-throughput two-hybrid systems [23]. Although such approaches significantly increase the rate at which interaction data is produced, they will require several years to produce full interaction maps for modest-sized organisms, whereas the "working draft" of the human genome has been available since June 2000. It is therefore enticing and promising to develop computational methods that could predict protein-protein interactions, be it in a rough and approximate manner. Ideally, the data produced by those high-throughput projects could suffice to develop prediction algorithms that could then be applied to genome sequence as fast as it is being released. More reasonably, the high-throughput projects themselves could benefit from predictions to speed up the discovery of interesting protein-protein interactions (see part 4).

In this paper we concentrate on a preliminary step to study protein-protein interactions in *Caenorhabditis elegans*. *C. elegans* is the first multi-cellular organism whose genome has been completely sequenced [5], as well as being a choice

O. Gascuel, M.-F. Sagot (Eds.): JOBIM 2000, LNCS 2066, pp. 135–146, 2001.

model organism for many functional genomics projects, from cDNA microarrays [9] to systematic knock-outs [10] and protein-protein interaction mapping. It is also a convenient model organism for classical genetic studies [24].

We first describe our efforts to set up WISTdb (Worm Interaction Sequence Tag database) [25], a database which gives access to interactions in *C. elegans*. These interactions are the result of the *C. elegans* interaction mapping project led by Marc Vidal at the Dana Farber Cancer Institute, Harvard Medical School, Boston, Massachusetts. Our goal consisted in setting up the informatics platform to annotate and store the interactions, and make them freely available through Internet. This has been implemented in the form of a database using the ACeDB database management system [21] and the AceBrowser [20] interface.

We then describe InterDB, a prediction-oriented database of protein-protein interactions, which we have built. The goal is to have access in a homogeneous way to as many known protein interactions present in available databases as possible, with the aim of predicting interactions in *C. elegans*.

We finally explain our current approaches to using this data. The aim is to extract rules that could explain the observed interactions of InterDB, and perhaps generalize to other unknown interactions. We report on our efforts at using an inductive logic programming technique, namely the Progol system [16], and another data mining technique based on association rules [2], [14].

2   The WISTdb Database

As said before, Marc Vidal's group at MGH is working on producing a protein interaction map for *C. elegans*, based on a high-throughput two-hybrid system. WISTdb is a database designed to store the data produced by this project. It is based on ACeDB, an object-oriented database management system developed initially to manage and distribute *C. elegans* genetic and genomic data. It also uses the Acembly sequence assembly and edition package [1], which functions on top of ACeDB. Both ACeDB and Acembly are freely available over the Internet. Since all the data currently available for *C. elegans* is distributed in the ACeDB format, the choice was a natural one. We first summarize the main ideas behind two-hybrid systems, then describe the database schema adopted for WISTdb and briefly discuss the algorithms involved.

The goal of conducting a two-hybrid experiment, or screen, is to identify proteins that physically interact with a given protein, called the bait. This bait can be expressed as a hybrid protein, fused to a specific DNA binding domain (coming from the yeast GAL4 transcription factor). Using a library of plasmids capable of expressing potential interactors fused to the GAL4 transcriptional activation domain, a custom yeast strain is co-transformed with two plasmids: the bait hybrid and one random plasmid from the library. The specifically engineered yeast cell is designed such that under certain conditions, its survival depends on the bait interacting with the unknown protein, thereby reconstituting the GAL4 activity. At this point, surviving yeast colonies contain a plasmid in which the cDNA insert codes for an interactor of the bait. Those interactors are called fishes. To identify them, they are amplified by PCR, then sequenced. The data produced is therefore a set of ABI automatic sequencer traces, for each bait screened. Our task is to align those traces on the *C. elegans* genomic sequence, so as to identify the fishes, and to store the newly discovered interactions in a database along with all relevant information. The main conceptual difficulties lay in the definition of a schema for the data. This is discussed below.

The data schema is derived from the schema distributed with the Acembly package. This software system is designed to handle ABI trace files and perform sequence assembly. It is well suited to the alignment of cDNA sequences on genomic sequence, and includes a versatile graphical interface to visualize and edit the traces. The new schema has been stripped of unnecessary classes and attributes, and enriched with the new classes IST and ISTScreen to describe the interactions found.

The main remaining standard classes are the following: locus, sequence, cdna_clone, and transcribed_gene. The locus class represents genetic loci, including gene names and other genetic information, as well as links to relevant sequence objects when available. The sequence class contains all nucleic acid sequences, be they genomic cosmids, expressed sequence tags (ESTs), predicted open reading frames, or ABI sequence traces. Cdna_clone objects store information on the clones from which the ABI traces come. Finally, when the traces are aligned on genomic sequence, they can be clustered into overlapping subsets of traces. The transcribed_gene objects are created by the system to reconstruct the genes corresponding to these clusters.

The new classes can be described as follows. An ISTScreen object is created for every gene used as a bait in a two-hybrid screen. It contains links to every IST object that represents an interaction uncovered by this screen. It also contains links to the gene (genetic locus) and sequence (physical locus) of the bait. An IST object is basically a link between two proteins that interact in the context of the two-hybrid system, a bait and a fish. It contains additional information such as the observed strength of the interaction, in terms of what two-hybrid phenotypes are observed. Both bait and fish may be referred to by gene name and/or sequence identifier. In fact, given a bait and a transcribed_gene, an IST linking them is generated if the transcribed_gene's construction relies on an ABI trace corresponding to a cdna_clone which scored positive in the two-hybrid screen with that bait.

The current version of WISTdb stores interaction data for 22 baits, therefore containing 22 ISTScreen objects. It also contains 1195 cdna_clone objects, which represent 117 different transcribed_gene objects, and correspond to 146 interactions (IST objects).

3   Construction of InterDB from Current Databases

The first step in studying protein-protein interactions is the construction of a database of such interactions. The main difficulties encountered are the scarcity of available data, and the problem of constructing an integrated database from several independent heterogeneous databases, as described below.

First, computationally useful protein interaction data is hard to find. Still today, biologists studying particular proteins search and identify interacting partners for their protein of interest. That information is eventually published, sometimes inconspicuously, along with other information on the protein. Reading biology papers can be a daunting task for non-specialists like ourselves, let alone reading thousands of them looking for specific data that rarely stands out. One approach to this problem is to apply natural language processing techniques, in order to extract protein-protein interaction information from the scientific literature. For example A. Valencia and colleagues, from Madrid University, have initiated such studies based on Medline abstracts (personal communication). But although this strategy appears promising, it can only give *predictions* of protein-protein interactions, since the natural language processing involved is not completely accurate. The only reasonable scenario for obtaining experimental data lay in either finding protein interaction databases compiled by third parties, or teaming up with functional genomics projects that produce such data. As said before, a collaboration was set up with the Vidal laboratory in Boston, which gives us access to the interactions that they detect. But more data was desirable. The following databases containing information on protein-protein interactions were found.

The DIP database (Database of Interacting Proteins, [12]) is a collection of interactions in a variety of organisms. Until recently, it used exclusively the PIR (Protein Information Resource, [26]) unique identifier to identify interacting proteins. The current release also provides SwissProt [3] identifiers when possible. DIP is freely available and downloadable, and contains around 2290 interactions at the time of writing.

Another source of interaction data is YPD (Yeast Protein Database, [27]). YPD is a general database on *Saccharomyces cerevisiae* proteins. It is the result of manual curation and annotation based on a review of the literature. YPD is a proprietary database, although access is freely granted to academic users. A lot of functional information is present, including protein-protein interactions, but it is in the form of free text in English, and scriptable access to YPD is forbidden. This would preclude using it in our context, but an agreement was negotiated with Proteome Inc., leading to our having access to a table containing all YPD protein-protein interactions. The table comprises 1115 interactions, and interacting partners are referred to by gene name.

Finally, FlyNets [19] contains to date 53 protein-protein interactions in the fly *D. melanogaster*. This database was considered but not used.

The second problem encountered is a classical one: starting with several heterogeneous databases, construct a single database integrating the knowledge stored in the initial ones. In our case, a unique framework was needed to identify proteins. The protein databases SwissProt and TrEMBL [3] were chosen. They are available as a non-redundant flat database, and respect a reasonably tight syntax. They contain links to corresponding PIR identifiers, as well as gene names.

For each source of protein-protein interaction data, a Perl script was written to parse the data files, and ace files were generated for all interactions in which both partners could be identified in SwissProt or TrEMBL. This guarantees the coherence of InterDB, but led to discarding some of the available interactions. Out of 1115 YPD interactions, 247 (22%) were discarded because at least one interactor could not be identified in SwissProt or TrEMBL, by the gene name given in YPD. Similarly, 1010 of the 2290 DIP interactions (44%) were discarded because they involved at least one protein whose PIR identifier was not referenced in SwissProt or TrEMBL.

A new database, called InterDB, was built to store the interaction data collected. It uses ACeDB as a database management system. The schema is designed to fulfill two main goals:

- the quick construction of the database from new releases of the source databases,
- the optimization of queries necessary to construct training sets of interactions, and to predict protein-protein interactions using predictive rules.

The schema contains the three main classes Protein, Interaction and Predictive_rule. Protein objects correspond to the SwissProt and TrEMBL entries. The information contained in the following SwissProt fields is stored: identification, including gene names and organisms concerned; database cross-references to PIR, ProSite and Pfam; keywords; and Sequence. Interaction objects are basically links between two protein objects. Predictive_rule objects are used to store predictive rules, as generated by the approaches described below.

InterDB contains to date 307199 protein objects, and 2245 interaction objects involving 1891 proteins. It should be noted that although only 1891 proteins are involved in interactions, InterDB must store all 307199 proteins from SwissProt and TrEMBL in order to identify interacting partners. 5% of the interactions are between *C. elegans* proteins, 75% between *S. cerevisiae* proteins, 10% between H. *sapiens* proteins, and the remaining 10% are spread over 40 various organisms. 45% of the 1891 proteins involved in at least one interaction are actually involved in two or more interactions.

4   Protein Interaction Prediction

As stated before, protein-protein interactions are fundamental to a wide range of biological processes, and several large-scale projects are under way to identify them experimentally. Yet the prediction of such interactions by computational methods remains a highly attractive goal, for reasons evoked earlier. Even if the predictions are not reliable enough to be useful to the final user, i.e. the biologist, they could still prove valuable, in the sense that they could guide the high-throughput projects to speed up the discovery of interesting protein-protein interactions.

Indeed, two-hybrid experiments can be performed in two different manners: the first approach is to screen against a library, as described in part 2; the second is to clone some genes into both DNA binding domain and transcriptional activation domain vectors, and to test all the cloned genes against one another. The main advantage of the first strategy is that each cloned gene can be tested and yield interesting results. Its main drawback is that every fished clone must be sequenced. The second approach requires that many genes be cloned beforehand, but the expensive sequencing step is avoided since the scientist knows what protein-protein interactions he is testing. In this context, cloning the genes in a favorable order can yield positive interactions rapidly. For example, if the aim is to map interactions concerning proteins involved in DNA repair, a possible strategy is to clone the fifty or so genes known to be involved in this process, and another few hundred genes suspected, or predicted, to interact with them.

It seems that bioinformatics has become interested in the question of predicting protein-protein interactions fairly recently. To our knowledge, two groups have published work in this direction [17]: Marcotte et al. [12] and independently Enright et al. [7] have developed methods to predict protein-protein interactions. They both rely on the hypothesis that when two proteins A and B are homologous to (a part of) a third protein in another organism, but are not homologous to each other, then they interact (the "fused domain" approach). Marcotte et al. also proposed a multiparadigm algorithm to predict functional links between S. cerevisiae proteins [13], which actually uses three prediction engines and two sources of experimental data. The first prediction engine is the fused domain algorithm discussed above. The second links proteins with related phylogenetic profiles [18], e.g. two proteins that have homologs in approximately the same subsets of genomes are predicted to interact. Finally, the third engine links proteins whose mRNA levels are correlated across various microarray experiments in S. cerevisiae [6]. This third method is not generalizable to other organisms without access to genome-wide expression data. However, the first two methods can be applied to any sequenced organism, and aim at predicting physical protein-protein interactions. But they are hypothesis-driven, meaning that their starting point is a biological hypothesis, which is implemented then validated.

5   An ILP Approach

Our method is different, in that we seek to discover rules that could explain protein-protein interactions, but don't have strong preconceptions as to what these rules should be. However, we do know that we need a formal language to express and manipulate the relevant biological background knowledge. The chosen formalism is first-order logic. The proposed predicates fall into two main categories: first, predicates that characterize individual proteins, specifically predicates that express the presence of a Pfam [4] or ProSite [8] domain in the protein considered, or its association with a keyword in SwissProt; second, predicates that express relationships between proteins or functional domains, for example predicates asserting that two proteins interact, or are homologs. We considered including predicates of the first category to use raw structural data from the PDB. Indeed, the information content of a 3D structure is clearly much higher than that of a Pfam or ProSite entry. In practice though, this idea is not implementable due to the scarcity of structural data concerning *C. elegans* (there are only 15 structures for partial or complete *C. elegans* proteins in the PDB as of November 2000). This language is used in the framework of inductive logic programming [15], which is an automated learning method inspired by logic programming and machine learning. More specifically, the powerful inductive logic programming system Progol [16] is used.

Progol generalizes a set of examples, i.e. positive instances of the predicate interaction, by generating Horn clauses from which these examples can be deduced. However, the induced Horn clauses can be specified to abide by user-defined rules, most notably the so-called mode declarations. Mode declarations permit to restrict the form that induced rules may take, for example by specifying whether the variables that appear in the atoms should be pre-bound or not. Allowing too many unbound variables, i.e. output variables, greatly increases the search space of inducible rules. Typically we decided, by using modes, to restrict the induced rules to the form:

```
interaction(P,Q):-
descriptor(P,D1)∧descriptor(P,D2)∧...∧descriptor(Q,D3)∧...,
```

where `descriptor(P,D)` is true if protein P is described by descriptor D, i.e. P contains Pfam or ProSite domain D, or is described by keyword D in SwissProt. Note that P and Q are always bound when they appear in the `descriptor` predicate: they were bound in `interaction(P,Q)`.

We tried several experiments with Progol. Unfortunately it appears that this approach is not suitable for such large amounts of data, both in terms of number of interactions and number of attributes. In fact, we had to restrict the size of our training sets to a maximum of 80 positive examples to avoid running out of memory and to stay within reasonable runtimes. Also, this method is particularly well adapted for dealing with highly structured and abundant background knowledge. Alas, the biological knowledge currently included in InterDB is mostly flat. For these reasons, we shifted towards a data mining technique supposedly better adapted to our data, namely association rules.

6   An Association Rule Approach

The idea behind the association rule data mining technique [2] [14] is the following. Given a boolean matrix where each line is a transaction and each column is an item, the goal is to find sets of items which are frequently present in the same transactions. From these *frequent itemsets*, one can then derive rules that link items. For example, if A and B are two items, and {A,B} is a frequent itemset, then A=¿B is derived, provided that its confidence (the ratio of the frequency of {A,B} over the frequency of {A}) is high enough. To apply this technique to our problem, we proceeded as follows.

The first task was to find a way of representing our data in an appropriate boolean matrix. The proteins corresponding to the 2245 interactions of InterDB are described by 497 Pfam domains, 406 ProSite domains, and 357 keywords, adding up to a total of 1260 descriptors. We therefore built a matrix comprising 2520 (1260×2) columns and 4490 (2245×2) lines. Each line represents an interaction, and for each line the first 1260 columns represent the presence of descriptors in the first protein, while the last 1260 columns correspond to the second protein. Since this introduces a dissymmetry not present in the inherently symmetrical "interaction" relation, we chose to enter interactions twice, once in each orientation, hence the 4490 lines. This choice means that for every frequent itemset, there is a dual itemset that appears with exactly the same frequency, and actually represents the same link between descriptors.

We then used a program, which implements the classic Apriori algorithm, to find frequent itemsets in this matrix. The frequency cutoff was set at 0.5%, meaning that an itemset had to be present in at least 0.5% of the lines to be considered frequent. This step produced 98391 frequent itemsets, along with their respective observed frequencies.

We finally applied a series of custom filters, designed to extract significant frequent itemsets from this list.

1. The first filter discards itemsets that concern only one protein. These itemsets actually correspond to linked descriptors within a single protein. For example, the well-known SH2 domain is represented by three different descriptors: one Pfam domain, one ProSite domain, and one SwissProt keyword. These three descriptors will therefore naturally form a frequent itemset. Although such itemsets are not deprived of meaning, they are not useful to predict interactions.

2. A second filter discards itemsets that contain specific user-specified descriptors. Typically, we don't want to consider itemsets containing the SwissProt keyword "hypothetical protein". Indeed, this keyword is obviously irrelevant to protein-protein interactions. Conceptually, these "bad" descriptors, which are actually mostly keywords, could have been eliminated before running the Apriori algorithm. But checking every descriptor beforehand would have been much more time-consuming than just looking at the descriptors that occur frequently and checking that they make sense with regards to protein-protein interactions. In a first approximation, we introduced an upper limit on the number of proteins described by each keyword to consider the keyword valid.

3. A third filter assigns a significance score to each itemset. The itemsets whose score is below a user-specified threshold are discarded. This score is defined as follows. Consider a frequent itemset I occurring with frequency F. We can write I=(D1,...,Dn,D'1,...,D'p), where D1...Dn are descriptors for protein 1 and D'1...D'p are descriptors for protein 2. The itemsets (D1,...,Dn) and (D'1,...,D'p) are also frequent, although they have been filtered out by step 1, and occur with frequencies F1 and F2 respectively. Supposing that these

two itemsets are independent, I is expected with a frequency F1xF2. We define the score for I as F/(F1xF2).

4. A final filter is then applied, in order to favor large itemsets vis-a-vis their subsets when the score penalty is not too heavy. Specifically, whenever two itemsets I1 and I2, whose scores are S1 and S2 respectively, are such that I1⊂I2, I1 is discarded if S1/S2 is smaller than a user-specified value.

To summarize, these filters generate a set of hopefully significant frequent itemsets, parameterized by three user-defined variables: a list of "bad" descriptors, a score cutoff, and a set to subset score ratio cutoff. Note that a frequent itemset (D1,...,Dn,D'1,...,D'p) can be interpreted as the rule:

interaction(P,Q) :-
descriptor(P,D1)∧...∧descriptor(P,Dn)∧descriptor(Q,D'1)∧...
∧descriptor(Q,D'p).

Based on this idea, a set of rules is generated for every triplet of parameter values. In practice, the filters are implemented in Perl, and a range of promising values has been determined for each parameter. The filters have been run for every combination of values in these ranges.

The next step involves validating the rules, and finding the optimal values for each parameter. A new set of experimentally determined protein-protein interactions has been produced recently by Anne Davy from CRBM, Montpellier in collaboration with Marc Vidal's laboratory. This test set TS contains 103 interactions, involving 81 proteins which play a role in the *C. elegans* proteasome. Since these interactions have not been used to produce current predictive models, they constitute a nice test set for these models.

In practice, the sets of rules are stored in InterDB and a Perl program has been developed to apply them to any input protein. The result is a list of potential interactors for that protein.

105 predictive models have been generated, using a promising range of values for each parameter. The upper limit on the number of proteins described by valid keywords has been set to 50, 100 and infinite (use all descriptors). The score cutoff has been set to 1, 2, 3, 5, 8, 10 and 50. The set to subset score ratio cutoff has been set to 1, 2, 3, 5 and infinite (keep all subsets). Using the most stringent values for these parameters, i.e. 50 for the upper limit, 50 for the score cutoff and 1 for the set to subset score ratio cutoff, we obtained 385 predictive rules. Using the most permissive values, i.e. infinite, 1 and infinite respectively, we obtained 83469 predictive rules.

We applied a representative subset of the predictive models to each TS protein, to obtain predicted interactions involving it. No interaction from TS was predicted successfully. This can be explained by the following observation: only 3 interactions from TS could possibly have been predicted by the most permissive model. We mean by this that those 3 interactions are the only ones in which each partner is described by at least one descriptor present in at least one predictive rule. We can propose two possible explanations for this. First, the initial frequency cutoff used in the Apriori algorithm seems too high, as only

79 descriptors out of 1260 are present in the predictive rules. Note that this is independent of the filters, since the most permissive model uses values which completely bypass filters 2 and 4, and reduce filter 3 to eliminating blatantly bad rules. Second, the information content of InterDB could be insufficient to produce pertinent rules for proteins from the proteasome.

7　Conclusion

Protein-protein interaction prediction is a difficult task, due to several reasons. First, it seems that there are not enough biological experiments to build training sets with enough coverage. Second, as always in bioinformatics, the data is never completely reliable. Third, counter-examples are not available, mostly due to the nature of the problem. This is particularly the case in a high-throughput setting, where it is vital to keep the generation of false positives as low as possible, therefore tolerating a higher rate of false negatives.

Three aspects of our work have been presented. We have first described the development of WISTdb, a platform designed to generate, store, annotate and make available the *C. elegans* protein-protein interaction data generated by the Vidal laboratory. Second, we have described InterDB, a prediction-oriented multi-organism protein interaction database. Finally, we have reported on our attempts to generate predictive rules for protein-protein interactions, using the Progol inductive logic programming system and an association rule data mining technique. We have not yet obtained satisfactory results with these approaches, perhaps for the reasons detailed above.

Work is under progress in two directions. On one hand, we plan to include richer and more structured biological knowledge in InterDB, from three sources. First, although the subcellular localization information from SwissProt is specified as free text, a close inspection reveals that most of the entries are chosen in a list of 66 localizations, which could be used as descriptors for our proteins. Second, we are replacing the Pfam and ProSite descriptors by InterPro descriptors (http://www.ebi.ac.uk/interpro/). InterPro is an integrated resource of protein families, domains and sites, and federates data from Pfam and ProSite, along with ProDom and Prints, two other databases with similar aims but different approaches. This resource will certainly provide more reliable information than the independent use of Pfam and ProSite, and also structures its entries by introducing relationships between them. Finally, we wish to incorporate data from the Gene Ontology Consortium (http://genome-www.stanford.edu/GO/), which features a hierarchical classification system for the functional annotation of proteins from Drosophila, Saccharomyces, Mus, Arabidopsis and Caenorhabditis. On the other hand, studies on fine-tuning the parameters used in the association rule approach are under way, especially by lowering the initial frequency threshold.

Acknowledgements. We wish to thank Jean-Francois Boulicaut and Baptiste Jeudy from INSA, Lyon for their help with the association rule data mining approach, and Anne Davy from CRBM-CNRS, Montpellier for sharing unpublished results concerning protein-protein interactions in the *C. elegans* proteasome.

References

1. The Acembly sequence assembly package,
   http://alpha.crbm.cnrs-mop.fr/acembly/
2. Agrawal R., Srikant R. (1994): Fast algorithms for mining association rules. *Proceedings of the 20th VLDB Conference*, 487-499
3. Bairoch A., Apweiler R. (1999): The SWISS-PROT protein sequence data bank and its supplement TrEMBL in 1999. *Nucleic Acids Research* **27(1)**, 49-54
4. A. Bateman, E. Birney, R. Durbin, S. Eddy, R.D. Finn, E.L. Sonnhammer(1999): Pfam 3.1: 1313 multiple alignments and profile HMMs match the majority of proteins. *Nucleic Acids Research*, 27(1), 260-262
5. The C. elegans Sequencing Consortium (1998), *Science* **282**, 2012-2018
6. M. Eisen, P. Spellman, P. Brown, D. Botstein (1998): Cluster analysis and display of genome-wide expression patterns. *Proc. Natl. Acad. Sci. USA* **95**, 14863-14868
7. A. Enright, I. Iliopoulos, N. Kyrpides, C. Ouzounis (1999): Protein interaction maps for complete genomes based on gene fusion events. *Nature* **402**, 86-90
8. K. Hofmann, P. Bucher, L. Falquet, A. Bairoch(1999): The PROSITE database, its status in 1999. *Nucleic Acids Research*, 27(1), 215-219
9. The Kim laboratory, http://cmgm.stanford.edu/~kimlab
10. The *C. elegans* Gene Knockout Consortium,
    http://www.cigenomics.bc.ca/elegans/
11. Lecrenier N., Foury F., Goffeau A. (1998): Two-hybrid systematic screening of the yeast proteome. *BioEssays*, **20**, 1-5
12. E. Marcotte, M. Pellegrini, H. Ng, D. Rice, T. Yeates, D. Eisenberg (1999): Detecting protein function and protein-protein interactions from genome sequences. *Science*, **285**, 751-753
13. Marcotte E., Pellegrini M., Thompson M., Yeates T., Eisenberg D. (1999): A combined algorithm for genome-wide prediction of protein function. *Nature* **402**, 83-86
14. Manilla H., Toivonen H., Verkamo A. (1994): Efficient algorithms for discovering association rules. *KDD-94: AAAI Workshop on Knowledge Discovery in Databases*
15. S. Muggleton, L. De Raedt(1994): Inductive logic programming: theory and methods. *Journal of logic programming*, 19,20:629-679
16. S. Muggleton(1995): Inverse entailment and Progol. *New generation computing*, 13, 245-286
17. A. Sali (1999): Functional links between proteins. *Nature* **402**, 23-26
18. Pellegrini M., Marcotte E., Thompson M., Eisenberg D., Yeates T. (1999): Assigning protein functions by comparative genome analysis: protein phylogenetic profiles. *Proc. Natl. Acad. Sci. USA* **96**, 4285-4288
19. C. Sanchez, C. Lachaize, F. Janody, B. Bellon, L. Röder, J. Euzenat, F. Rechenmann, B. Jacq(1999): Grasping at molecular interactions and genetic networks in Drosophila melanogaster using FlyNets, an internet database. *Nucleic Acids Research* **27(1)**, 89-94
20. L. Stein, J. Thierry-Mieg (1999): Scriptable Access to the Caenorhabditis elegans Genome Sequence and other Acedb Databases. *Genome Research* **8(12)**:1308-1315
21. J. Thierry-Mieg, D. Thierry-Mieg, L. Stein (1999): ACEDB: The ACE database manager. In S. Letovsky (ed.): *Bioinformatics, Databases and Systems*, Kluwer Academic Publishers, 265-278

22. Uetz *et al.* (2000): A comprehensive analysis of protein-protein interactions in Saccharomyces cerevisiae. *Nature,* **403**, 623-627

23. M. Vidal, P. Legrain (1999): Yeast forward and reverse 'n'-hybrid systems. *Nucleic Acids Research* **27(4)**, 919-929

24. A. Walhout, H. Endoh, N. Thierry-Mieg, W. Wong, M. Vidal (1999): A model of elegance. *American Journal of Human Genetics* **63(4)**:955-61

25. A. Walhout, R. Sordella, X. Lu, J. Hartley, G. Temple, M. Brasch, N. Thierry-Mieg, M. Vidal (2000): Protein interaction mapping in C. elegans using proteins involved in vulval development. *Science,* **287**, 116-122

26. Winona C. Barker, John S. Garavelli, Peter B. McGarvey, Christopher R. Marzec, Bruce C. Orcutt, Geetha Y. Srinivasarao, Lai-Su L. Yeh, Robert S. Ledley, Hans-Werner Mewes, Friedhelm Pfeiffer, Akira Tsugita and Cathy Wu (1999): The PIR-International Protein Sequence Database. *Nucleic Acids Research* **27(1)**: 39-43

27. The Yeast Protein Database, http://www.proteome.com/

Application of Regulatory Sequence Analysis and Metabolic Network Analysis to the Interpretation of Gene Expression Data

Jacques van Helden[1,2], David Gilbert[2,3], Lorenz Wernisch[2], Michael Schroeder[3], and Shoshana Wodak[1,2]

[1] SCMBB. Université Libre de Bruxelles
CP160/16. 50 av F.D. Roosevelt. B-1050 Bruxelles. Belgique.
{jvanheld,shosh}@ucmb.ulb.ac.be
[2] European Bioinformatics Institute.
Genome Campus - Hinxton Cambridge CB10 1SD - UK.
{jvanheld,drg,lorenz,shosh}@ebi.ac.uk
[3] Department of Computing, City University.
Northampton Square, London EC1V 0HB, UK.
{drg,msch}@soi.city.ac.uk

Abstract. We present two complementary approaches for the interpretation of clusters of co-regulated genes, such as those obtained from DNA chips and related methods. Starting from a cluster of genes with similar expression profiles, two basic questions can be asked:
1. Which mechanism is responsible for the coordinated transcriptional response of the genes? This question is approached by extracting motifs that are shared between the upstream sequences of these genes. The motifs extracted are putative cis-acting regulatory elements.
2. What is the physiological meaning for the cell to express together these genes? One way to answer the question is to search for potential metabolic pathways that could be catalyzed by the products of the genes. This can be done by selecting the genes from the cluster that code for enzymes, and trying to assemble the catalyzed reactions to form metabolic pathways.
We present tools to answer these two questions, and we illustrate their use with selected examples in the yeast *Saccharomyces cerevisiae*. The tools are available on the web (http://ucmb.ulb.ac.be/bioinformatics/rsa-tools/;
http://www.ebi.ac.uk/research/pfbp/; http://www.soi.city.ac.uk/~msch/).

1 Introduction

DNA chips [1-3] and related techniques permit the measurement of the transcriptional response of all the genes of an organism to a controlled stimulus (presence/absence of metabolites, action of a drug, temperature, ...) or to a genetic modification (deletion or over-expression of a selected gene). Results of several experiments are combined into a multivariate table, summarizing the response of all the genes of an organism to a

O. Gascuel, M.-F. Sagot (Eds.): JOBIM 2000, LNCS 2066, pp. 147–163, 2001.

variety of conditions. Genes can then be clustered on the basis of similarities in their expression profiles. Different approaches have been used for this purpose: hierarchical clustering [3], self-organizing maps [4], k-means [5]. Once such clusters have been obtained, two complementary questions can be asked (Fig. 1).

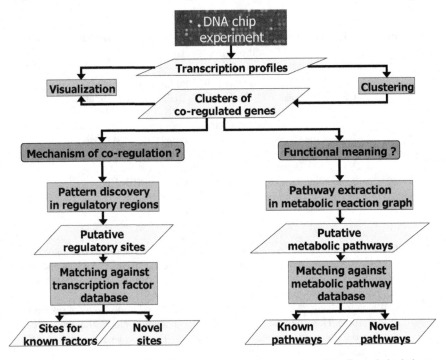

Fig. 1. Flow chart of the approaches for interpreting gene expression data. Round shaded rectangles represent the questions, shaded rectangles the processes, and white parallelograms data

(1) Which mechanism could ensure the coordinated transcriptional response of the genes belonging to a given cluster? Transcriptional regulation is mediated by a class of proteins, called transcription factors, which bind specific DNA motifs, called cis-acting elements, and interact with RNA polymerase to enhance (activation) or reduce (repression) the expression of neighboring genes. The answer to this question amounts to a search for transcription factors that might act simultaneously on the genes belonging to the cluster. There is no obvious way to predict directly which proteins act in trans on a set of genes, but the problem can be addressed indirectly by first predicting which cis-acting elements might be involved, and then looking for transcription factors that might bind these cis-acting elements. The approach consists in analyzing upstream sequences to discover shared motifs, which could correspond to regulatory elements. Candidate cis-acting elements can then be matched against databases of known binding sites [6,7], and/or tested experimentally.

(2) Which biological function requires a coordinated expression of the genes belonging to the cluster under consideration? It is usual that genes involved in a common

process are co-regulated, ensuring the presence of all the necessary proteins. The question amounts thus to search for processes in which most genes of the cluster might be involved. One simple approach is to match the set of genes against a database of gene/protein function [8-9]. This would however restrict the possible answers to processes/pathways that have already been previously characterized and are stored in the database. A more flexible approach is to try to identify reactions that could be catalyzed by the gene products, and to interconnect these reactions in any possible way to generate potential metabolic pathways. The pathways assembled by this way can then be matched against metabolic pathway databases. Part of these pathways will correspond to previously described pathways, whereas in other cases one should be able to discover novel pathways.

This paper is a mini-review of our recent work on several aspects related to the interpretation of gene expression data. We illustrate the different questions that can be addressed on the basis of a selected study case, and discuss some critical issues for obtaining suitable results. We refer to previously published work for a detailed description of the statistical and algorithmic aspects, which would go beyond the scope of this review.

2   Gene Expression Data: A Study Case

To illustrate our purpose, we selected an example of gene expression data from the literature. Spellman and co-workers used the DNA chip technology to detect yeast genes that are involved in cell cycle [10]. These authors measured the level of expression of all 6000 yeast genes at different time points during the cell cycle, and selected those showing periodic fluctuations. The 800 selected genes were then clustered according to similarities in their expression profiles. Some of the clusters obtained were clearly associated to well defined cellular processes associated to the cell cycle. An unexpected cluster was also isolated, mostly made of genes involved in methionine biosynthesis. We will use this MET cluster as study case throughout the following chapters.

3   Visualization

The development of flexible and intuitive visualization tools is an important requirement for the interpretation of gene expression data (Fig. 1). One popular approach has been to apply hierarchical clustering and to display the profiles of expression in parallel with the dendrogram [11]. We are currently working on complementary approaches, which would provide a direct representation of the functional distances between genes [12]. This is illustrated in Fig. 2, which shows a mapping of the 800 genes from Spellman's experiment on a Euclidian space. Each dot represents a single gene. Coordinates were assigned so that the distances between dots reflect the dissimilarities between gene expression profiles. Noticeably, genes are grossly aligned along a ring, which is probably the most direct way to represent cell cycle. In particular, the center of the ring is avoided, and most genes align on the periphery, whereas random data would occupy the center as well as the periphery (not shown). Genes with synchronous fluctuations of expression appear in the same angle of the circle. The MET cluster mentioned above is highlighted.

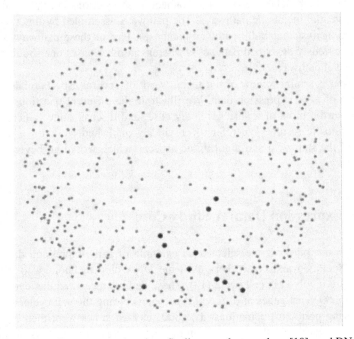

Fig. 2. Visualization of gene expression data. Spellman and co-workers [10] used DNA chips to measure the level of expression of all yeast genes during cell cycle, and isolated 800 gene showing periodical fluctuations. In the original paper, genes are displayed on a tree. In the alternative representation shown above, genes are mapped on an Euclidian space. Each dot represents a gene. Coordinates are assigned so that the distance between two dots reflects the distance between the corresponding gene expression profiles. In the case of cell-cycle regulated genes, most dots align on the periphery of a ring, and the center is avoided (which would not be expected from random data). Genes having a synchronized peak of expression are located in the same angle of the ring. The genes from the MET family, our study case, are highlighted. Note that the visualization programs support true colors, and would allow to discriminate several clusters on the same image [12]. The present image is in grayscale due to publication restrictions

4 Regulatory Sequence Analysis

Features of Cis-Regulatory Elements

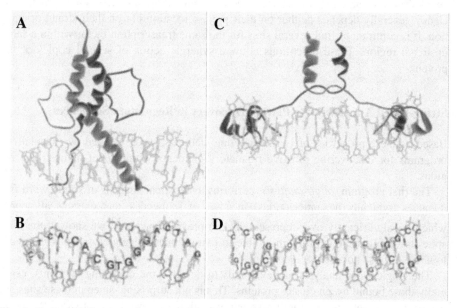

Fig. 3. Structural features of transcription factor-DNA interfaces. **A.** Pho4p-DNA complex. Notice the two protein monomers acting on DNA like tweezers, and the restricted number of adjacent nucleotides that enter into direct contact with the protein. **B.** Pho4p binding site. The central oligonucleotide shown in larger characters (CACGTGGG) is conserved among Pho4p between sites. **C.** Gal4p-DNA complex. Gal4p belongs to the Zn cluster family of transcription factors. Notice the pair of trinucleotides entering into direct contact with the protein, separated by an intermediate region. **D.** Gal4p binding site. In contrast with Pho4p, the conserved region is not an oligonucleotide but a pair of trinucleotides (GGC-GGC) separated by a 11bp region of weakly conserved bases

Transcription factors bind to short stretches of DNA, whose sequence is highly specific for each particular transcription factor. The specificity of the binding site is determined by the structure of the protein domain that enters into direct contact with DNA. Most sites described can be regrouped into two classes. The first class consists of a short sequence of highly conserved adjacent nucleotides (typically 6 base pairs), surrounded by a few partly conserved nucleotides (Fig. 3A,B). This kind of site is associated to proteins containing different types of DNA-binding domains: Zn finger, homeodomain, leucine zipper, basic Helix-Loop-Helix. Another class of binding sites consists of a pair of very short oligonucleotides (typically 3 base pairs each), separated by a region of fixed width but variable content (Fig. 3C,D). These sites are typical for factors containing a Zn cluster domain (found in fungi) or a Helix-Turn-Helix domain (the most frequent DNA-binding domain in prokaryotes). Many of these sites show an internal symmetry (tandem repeat or reverse palindrome) due to the fact that the transcription factor binds DNA in a homodimeric form, each monomer entering into direct contact with one trinucleotide.

In the yeast *Saccharomyces cerevisiae*, cis-acting elements are found upstream the genes they control, within a range of 800 base pairs from the start codon. Their efficiency generally depends neither on their precise location nor on their strand orientation. It is common to find several sites for the same transcription factor within a same upstream region. These repetitions allow a synergic action of several copies of the protein.

String-Based Approaches to Pattern Discovery in Regulatory Sequences

Based on these properties of yeast cis-acting elements, we developed two specialized programs for discovering putative regulatory elements within a set of upstream regions.

The first program, *oligo-analysis*, performs a statistical analysis of single word frequencies (typically hexanucleotides) in a set of sequences, and extracts all words which are significantly over-represented. In a previous paper [13] we showed that, despite its simplicity, this program is able to extract many regulatory sites with a very low rate of false positives.

The single-word analysis however fails to detect some cis-acting elements, especially those bound by Zn cluster proteins. This is not surprising, since in these sites the conserved nucleotides are shared between two sub-sites separated by a variable region. Consequently, we developed a complementary program, *dyad-analysis* [14], which performs a statistical analysis of all possible pairs of shorter words (typically trinucleotides) with different spacing values (between 0 and 20). This program is very efficient, not only for the detection of sites bound by Zn cluster and HTH proteins, but also for sites recognized by other classes of transcriptional factors.

Criteria for the Statistical Analysis of Words and Dyads

The efficiency of the above mentioned programs crucially depend on the choice of appropriate parameters.

Selection of Non-redundant Upstream Sequences

Before envisaging any analysis of upstream sequences, it is essential to avoid redundancy in the data set. Indeed, the applicability of the statistical tests relies on the mutual independency of the sequences. Two sources of redundancy can be identified.

- Recent duplication of a gene, together with its upstream region. Such duplications are particularly common in yeast telomeric regions.

- Intergenic region shared between two divergently transcribed neighbor genes. If both genes belong to the same cluster, the intermediate upstream region will be included twice in the sequence set.

The data set has thus to be purged by discarding sequences that show a high similarity with the direct or reverse strand of any other sequence of the set.

How to Count Word Occurrences?

Should words be counted on a single or both strands? How to treat overlapping occurrences of a same word? The choice of the counting mode depends on the expected characteristics of regulatory sites, which can differ between organisms. In the case of yeast, best results are obtained by counting occurrences on both strands and without overlap.

Estimation of Expected Word Frequencies

The simplest approach would be to consider all words as equiprobable. This would however provide bad results, due to the high frequency of A and T nucleotides in yeast sequences. This bias can be corrected by calculating expected word frequencies on the basis of nucleotide-specific frequencies (residue frequencies). This correction already provides better results, but still returns many false signals, mainly AT-rich sequences, due to a preferential aggregation of A and T nucleotides in yeast non-coding sequences. The best approach consists in calculating pre-defined tables of expected word frequencies (background frequencies), based on the whole set of yeast non-coding sequences.

How to Compare Expected and Observed Frequencies?

Several statistics have been envisaged for detecting over-represented words in DNA sequences: observed/expected ratio [15], log likelihood [16], Poisson distribution [17], Z-values [18], binomial distribution [13,14]. The observed/expected ratio has to be avoided, because it is strongly biased in favour of patterns with low expected frequencies. The log-likelihood introduces a correction for this bias, but is not easy to convert to probabilistic values. Z-scores rely on an assumption of normality of the distribution of occurrences, which is only verified when sequence length tends towards infinite. For small sequence sets (a few thousands of base pairs) such as families of upstream sequences from co-regulated genes, the most appropriate statistics are Poisson and binomial.

How to Select the Threshold of Probability?

Analyzing a single sequence set involves a comparison between observed and expected frequencies for several thousands of words. For example, there are 4,096 possible hexanucleotides. This means that with a probability threshold of 0.01, around 40 words would still be selected from any random sequence. According to the Bonferoni rule, these false positive can be avoided by lowering the threshold of probability to a value lower than $1/4096 = 0.00025$. For heptanucleotides, the threshold should be lowered to 1/16,564. Thus, the threshold value has to be adapted to the number of words taken into consideration, which itself depends on the word length.

Significance Index

We defined a significance index [13,14] which provides an intuitive way to evaluate the degree of over-representation.

$$sig = -\log_{10}\left[P(occ \quad n)*D\right]$$

Where D is the number of possible patterns, and $P(occ \geq n)$ the probability to observe n or more occurrences for the word considered. This probability can be calculated with the binomial, Poisson or normal forumula, as described above.

The significance index takes into consideration the effects mentioned above, including the reduction of significance when the number of possible patterns increase. The index can take positive or negative values. The interpretation is fairly intuitive. Positive values indicate over-representation. In random sequence sets, one expects to find no more than one pattern with a positive value, independently of the conditions such as sequence length, number of sequences, pattern length, ... A value higher than 1 is expected every 10 sequence sets, a value higher than 2 every 100 sequence sets, and more generally a value higher than s is expected every 10^s sequence sets. The index applies to spaced dyads [14] as well as single oligonucleotides [13].

How to Select Word Length?

Small words (di- to tetra-nucleotides) present a marked bias from theoretical distributions, and Poisson/binomial statistics are thus inappropriate. On the other side, analyzing too large words (octa-, nona-nucleotides) would prevent any of them to be detected as significant. Practically, we observed that hexanucleotide analysis provides excellent results in most cases for yeast sequences [13].

Even when restricting the analysis to hexanucleotides, larger patterns can nevertheless be detected, by assembling strongly overlapping hexanucleotides.

All the statistical considerations above are easily extended to the analysis of spaced dyads [14].

Evaluation of String-Based Approaches with Known Regulons

In order to evaluate the above methods, sets of genes were collected for which the transcription factor was already known. The programs were fed with the upstream sequences, and the significant patterns were compared with the expected binding sites [13, 14]. Generally, the number of significant words/dyads is restricted to a dozen per gene set. Some of these selected words/dyads strongly overlap with each other, and can be combined (using a custom fragment assembly program called *pattern-assembly*) to form a larger pattern. Pattern assembly also allows to describe, to some extent, the partial degeneracy of some binding sites. Indeed, it is frequent to detect several patterns that differ by a single substitution, and correspond to variants recognized by the same transcription factor. Patterns with a high significance index are always associated with known transcription factors. Some additional patterns appear that might be associated with novel transcription factors.

Application to Clusters Obtained from Microarray and Related Technologies

After having evaluated the programs with known regulons, we applied the same string-based approaches to extract putative regulatory elements from clusters of genes resulting from DNA chip experiments. We published elsewhere [14] an analysis of families of cell cycle regulated genes defined by Spellman [10]. We show here in more details the results obtained with the MET family (Table 1). All pairs of trinucleotides separated by spacing between 0 and 20 were anaylzed. The significant patterns form three groups of overlapping words, that can be assembled into 3 larger patterns. One additional isolated dyad is selected.

The first group of words corresponds to the binding site of the Met4p/Met28p/Cbf1p complex. The second group corresponds to Met31p and Met32p binding sites. All these transcription factors are known to act cooperatively to activate transcription of genes related to methionine metabolism. The highest score within each group is highlighted in bold. The pattern selected with the highest score generally corresponds to nucleotides that enter into direct contact with the transcription factor.

It is not possible to evaluate the efficiency of the programs on families obtained from DNA chip experiments with the same precision as was done with known regulons, since the transcription factors are usually not known. However, we observed that the same kind of result is generally obtained: a very restricted number of words/ dyads are selected as significant. For some families, patterns are selected with a very high significance index, suggesting a very likely putative regulatory element. In other families, the patterns selected have a lower significance index. This is often the case for very small families (less than 5 genes), and results from a reduction of the signal-to-noise ratio. On the other extreme, analyzing too large gene clusters (> 50) reduces the sensitivity of the programs. The reason is that the larger clusters are less likely to be regulated by a single factor, and might contain a mixture of different signals. The effect of mixing together sequence that contains a given signal with sequences that do not contain it is also to reduce the signal-to-noise ratio. The programs are able, to some extent, to extract multiple signals from a single analysis, but the highest efficiency is clearly obtained by selecting clusters of genes that are likely to be all regulated by the same transcription factor. The choice of the clustering method is thus crucial.

5   Metabolic Network Analysis

We focus now on the second question, namely the functional interpretation of clusters of co-regulated genes.

Table 1. Patterns extracted by dyad-analysis with the MET family. Legends: **obs occ**: observed occurrences; **exp occ**: expected occurrences; **proba**: binomial probability; **sig**: significance index. All patterns with significance value higher than 1 were selected. Some patterns can be grouped together on the basis of sequence similarities, and assembled into larger patterns (contigs). The first group corresponds to the sequence recognized by the protein complex Met4p/Met28p/Cbf1p. The second group describes the site bound by Met31p and its homologue Met32p. These factors are those known to regulate methionine biosynthesis in the yeast *Saccharomyces cerevisiae*. To our knowledge, the third group and the isolated patterns do not show any obvious similarity to known binding sites, and could reveal new regulatory patterns

	pattern	reverse complementary	obs occ	exp occ	proba	sig
group 1	GTC..GTG..	..CAC..GAC	17	2.61	3.60E-09	3.8
	.TCA.GTG..	..CAC.TGA.	23	5.12	8.50E-09	3.4
	.TCACGT...	...ACGTGA.	21	4.75	4.60E-08	2.7
	..CACGTG..	..CACGTG..	**38**	**3.37**	**0**	**20**
	..CAC.TGA.	.TCA.GTG..	23	5.12	8.50E-09	3.4
	..CAC..GAC	GTC..GTG..	17	2.61	3.60E-09	3.8
	...ACGTGA.	.TCACGT...	21	4.75	4.60E-08	2.7
assembly	GTCACGTGAC	GTCACGTGAC				
group 2	CGCCAC....	....GTGGCG	14	2.32	2.00E-07	2.1
	.GCCACA...	...TGTGGC.	21	4.06	3.30E-09	3.8
	..CCA..GTT	AAC..TGG..	23	5.86	9.10E-08	2.4
	..CCACAG..	..CTGTGG..	**24**	**3.53**	**2.30E-12**	**7**
	..CCA.AGT.	.ACT.TGG..	21	4.59	2.60E-08	2.9
	...CACAGT.	.ACTGTG...	24	5.41	5.20E-09	3.6
	...CAC.GTT	AAC.GTG...	24	5.79	1.90E-08	3.1
assembly	CGCCACAGTT	AACTGTGGCG				
group 3	ACC..............TGG.	.CCA..............GGT	15	2.9	5.10E-07	1.7
	.CCA..............GGT	ACC..............TGG.	15	2.9	5.10E-07	1.7
	.CCA..............TGG.	.CCA..............TGG.	22	3.16	1.10E-06	1.3
assembly	ACCA..............TGGT	ACCA..............TGGT				
isolated	CAG...TGG	CCA...CTG	17	3.12	4.60E-08	2.7

Representating Metabolic Pathways as Graphs

The set of all possible metabolic reactions can be seen as a graph, with two types of nodes (metabolites and reactions respectively). Arcs represent substrate-reaction and reaction-product relationships. A graph containing all known metabolic reactions would include of the order of 10^4 nodes and as many arcs. The connectivity is very high for some particular compounds (ATP, Adenosyl-Methionine), but besides these "pool metabolites", the vast majority of compounds are involved in a very limited number of reactions. Reactions have between 1 and 6 substrates (2 on average) and as many products. The complexity of such a graph is huge and the number of possible pathways is virtually infinite.

However, only a very restricted number of these possible pathways are effectively followed in living organisms. For instance, the database EcoCyc, which holds the most comprehensive information about *Escherichia coli* metabolism, only contains 159 distinct pathways. *E. coli* has been, for several decades, the preferred model organism for biochemists, and even though some parts of its metabolism certainly remain to be discovered, the number of pathways is not expected to increase significantly for this organism.

Metabolic Pathway Discovery

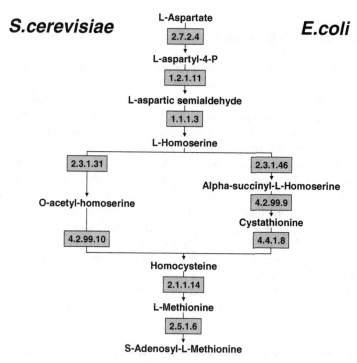

Fig. 4. Alternative pathways for methionine biosynthesis in *E. coli* and *S. cerevisiae*

Many pathways however remain be discovered in other organisms. Indeed, it is common to observe that different organisms follow distinct pathways for the biosynthesis or degradation of the same molecule. For example, in *E. coli*, methionine is synthesized in 7 steps from aspartate, whereas the yeast *Saccharomyces cerevisiae* performs this transformation in 6 steps, 4 of which are common with *E. coli* (Fig. 4). The case of lysine is more extreme, since *E. coli* and *S. cerevisiae* follow completely different pathways to synthesize this metabolite.

In addition, many parts of the metabolism remain largely unexplored, for example the mechanisms of toxic molecule degradation or resistance to extreme conditions observed in some bacteria.

In summary, among all the pathways that could be followed in the graph of metabolic reactions, only a very restricted fraction corresponds to already described pathways. Another part corresponds to pathways that are not yet described but might appear to be effectively used by some organisms in response to some conditions. Finally, a vast majority of these pathways might be devoid of any biological relevance. As illustrated below, measuring the transcriptional response of all the genes of an organism could be one way to select those pathways that are most likely to correspond to biological processes.

Metabolism and Gene Expression

Living organisms can rapidly modify their internal concentration of small molecules (metabolites) via enzymatic catalysis. Controlling metabolite fluxes is essential to cell viability, in that it allows the cell to maintain biochemical compounds in stationary concentrations (homeostasis), in spite of fluctuations of their external availability and internal consumption rates. Several molecular mechanisms are involved in metabolic regulation. Enzymes and transporters are regulated at different levels: transcription rate, RNA stability, translation rate, protein activity, intracellular location, protein degradation. Several of these mechanisms can be combined for the control of the same metabolic pathway. Enzymes and transporters participating in a common metabolic pathway are often co-regulated at the transcriptional level. Thus, when the culture medium is modified by depleting (or adding) a given metabolite, it is expected that the genes that participate in the biosynthesis (or degradation) of the molecule will respond at the transcriptional level. DNA chip and related technologies can be used to unravel the set of genes that respond to a given perturbation of the external conditions (addition/removal of a metabolite) in a given organism. The question is thus to discover, from this set of genes, which particular pathway could be catalyzed.

Applying Graph Analysis for a Functional Interpretation of Gene Expression Data

The first step is to select, among the set of co-regulated genes, those that code for enzymes, and identify the reactions they could catalyze. These reactions correspond to a subset of nodes in the graph of all possible metabolic reactions (Fig. 5A). The method consists in trying to interconnect all these reactions in a meaningful way (Fig. 5B), in order to extract a sub-graph (Fig. 5C) corresponding to one or several putative metabolic pathways (Fig. 5D). The algorithms for subgraph extraction and maximal path enumeration will be described elsewhere (*in prep.*), and we will only summarize their principle.

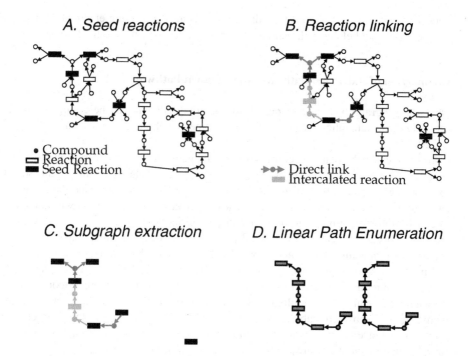

Fig. 5. Conceptual schema of the subgraph extraction. **A:** graph representation of metabolic pathways. Two types of nodes are used to represent metabolites (circles) and reactions (rectangles) respectively. Arcs represent the relationships between reactions and their substrates and products. Filled rectangles represent seed reactions, i.e. those that are catalyzed by genes belonging to the co-regulated cluster. **B:** A direct link (dark gray) can be established between two reactions when the first one produces a metabolite that is used as substrate by the second one. Optionally, one can decide to allow intercalating reactions (ligt gray) that were not part of the initial seeds, if this improves the connectivity. **C:** Connected components are then extracted from the graph. The extracted subgraph represents a metabolic pathway that is potentially catalyzed by the cluster of co-regulated genes. **D:** When the subgraph contains branches, it can be decomposed into non-redundant elmentary paths, which highlight potential endpoints of the metabolic pathways

The simplest way to interconnect reactions is to identify compounds that are produced by one reaction and used as substrate by another one. In a second step, linking can be improved by intercalating reactions that were not part of the initial set. Several reasons could be invoked to justify such an intercalation. Firstly, some genes could be involved in the metabolic pathway without being regulated at the transcriptional level. Secondly, microarray technologies are still limited in reproducibility and some regulations might have escaped detection. A third possibility would be that the gene is present on the chip and its expression level has been measured correctly, but this gene has not been annotated as an enzyme yet. Indeed, for newly sequenced genomes, gene function is usually predicted by sequence similarities, and many genes remain unannotated. In such a case, the best candidates to ensure the missing enzymatic catalysis are the genes that belong to the initial cluster itself, but have no assigned function yet.

Comparison of Extracted Pathways with Known Pathways

Once the subgraph has been extracted, the putative pathway can be compared to the set of known metabolic pathways stored in some metabolic pathway database [8, 9, 19, 20].

In some cases the pathway extracted from the gene cluster will correspond to some previously characterized pathway. For such cases, a simple matching of the set of reactions against a database of metabolic pathways would have provided the same answer. In other cases, one might observe only a partial match with a known pathways. The subgraph extraction might thus reveal an alternative to the pathway followed in the model organism. The method could be applied to study the metabolism of newly sequenced organisms, whose metabolism has been poorly characterized.

Finally, in some cases, one should be able to extract completely novel pathways. The co-regulation of the enzyme-coding genes would provide a good support to indicate that this pathway is biologically relevant. An interesting field of application would be to discover metabolic pathways involved in largely unexplored processes, such as resistance to toxic compounds or extreme conditions. Another application is to reveal which metabolic pathways are affected by a new drug.

Application of Pathway Analysis to the Study Case

We applied the above procedure to the 20 genes belonging to the MET cluster defined by Spellman and co-workers. Seven of these genes code for enzymes, which can catalyze 8 distinct reactions. Subgraph extraction and maximal path enumeration resulted in a linear pathway including 6 of the initial reactions (Fig. 6). In this case, the linear path was obtained without intercalating any reaction that was not part of the initial set.

The pathway shows partial matches with two distinct metabolic pathways: the 4 initial steps match the sulfur assimilation pathway, and perform a progressive reduction of sulfate into sulfide. The two last steps match the methionine biosynthesis pathway, and correspond to the incorporation of sulfur into homocysteine, and the transformation of the latter into methionine. Sulfur assimilation and methionine biosynthesis are intrinsically related in the yeast Saccharomyces cerevisiae, since in this organism sulfur amino acids are all derived from the methionine biosynthesis pathway (this differs from Escherichia coli, where sulfur is incorporated into cysteine and then transferred to methionine). It makes thus sense to have a coordinated transcriptional regulation of all the metabolic steps from sulfate to methionine. Indeed, these genes are all known targets of the methionine-regulating transcription factors described above [21].

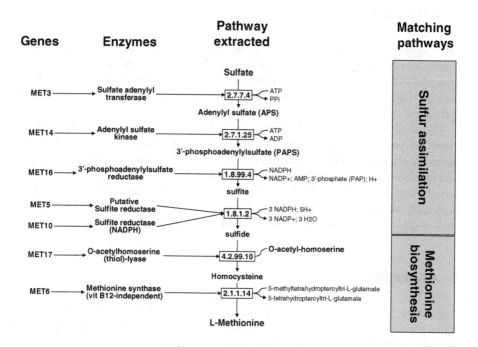

Fig. 6. Result obtained by pathway analysis with the cell-cycle regulated gene cluster MET from Spellman [10]. Six of the reactions potentially catalyzed by enzymes-coding genes from the cluster can be assembled into a single linear pathway, without need to intercalate any additional reaction. The pathway extracted is the way used by yeast to incorporate sulfur into amino acids, by reduction of sulfate into sulfide, which is incorporated in homocyteine. This pathway matches two distinct pathways from the database: the first 4 steps correspond to sulfur assimilation, whereas the two last steps are part of the methionine biosynthetic pathway

In summary, starting from an unordered set of reactions, the program was able to build a linear metabolic pathway, which correspond to our expectation for the study case. In this particular case, the pathway was already well characterized and a similar result would have been obtained by matching the seed reactions against a database of metabolic pathways like KEGG. However, since this pathway was re-discovered by the program without any a priori information about how reactions do assemble into pathways in the yeast (the matching with known pathways was only done a posteriori), one can hope that the same method will also provide a means of discovering novel pathway. We are currently optimizing the program and evaluating it performances in different conditions, on the basis of well characterized pathways. The optimized program will then be used to provide an interpretation of gene expression data in terms of metabolic pathways.

6 Conclusions

In the context of genomic approaches, coding sequence analysis is often insufficient to systematically assign a function to each gene. The function depends not only on the structure of the encoded protein, but also on the context in which this protein exerts its activity. Functional predictions thus require the integration of different levels of information.

The possibility to measure the transcriptional response at a genome scale offers exciting perspectives for the discovery of gene function, taking into account the ways genes are associated in functional clusters. By combining regulatory sequence analysis and metabolic pathway analysis, one could obtain two independent and complementary sources of information for these clusters of co-regulated genes. The same methods also apply to clusters of genes obtained from other functional genomics approaches, such as phylogenetic profiles [22] and gene fusion/fission analysis [23-25].

7 Availability

Regulatory Sequence Analysis tools are available on the web [26] at the URL http://ucmb.ulb.ac.be/bioinformatics/rsa-tools/. The home page for the EBI project of database on Protein Function and Biochemical Pathways is at http://www.ebi.ac.uk/research/pfbp/. A prototype version of the patwhay analysis tools can be accessed from this site. A prototype version of the visualization tools is available at http://www.soi.city.ac.uk/~msch/.

Acknowledgements. Jacques van Helden was funded by the European Commission Contract N0: QLRI-CT-1999-01333.

References

[1] DeRisi, J.L., Iyer, V.R. & Brown, P.O. Exploring the metabolic and genetic control of gene expression on a genomic scale. *Science* **278**, 680-6 (1997).

[2] Brown, P.O. & Botstein, D. Exploring the new world of the genome with DNA microarrays. *Nat Genet* **21**, 33-7 (1999).

[3] Eisen, M.B. & Brown, P.O. DNA arrays for analysis of gene expression. *Methods Enzymol* **303**, 179-205 (1999).

[4] Tamayo, P. *et al.* Interpreting patterns of gene expression with self-organizing maps: methods and application to hematopoietic differentiation. *Proc Natl Acad Sci U S A* **96**, 2907-12 (1999).

[5] Vilo, J., Brazma, A., Jonassen, I. & Ukkonen, E. Mining for Putative Regulatory Elements in the Yeast Genome Using Gene Expression Data. *ISMB* (2000).

[6] Wingender, E. *et al.* TRANSFAC: an integrated system for gene expression regulation. *Nucleic Acids Res* **28**, 316-319 (2000).

[7] Salgado, H. *et al.* RegulonDB (version 3.0): transcriptional regulation and operon organization in Escherichia coli K-12. *Nucleic Acids Res* **28**, 65-67 (2000).

[8] van Helden, J. *et al.* From molecular activities and processes to biological function. *Briefings in Bioinformatics* **in press**(2001).

[9] van Helden, J. *et al.* Representing and analysing molecular and cellular function using the computer. *Biol Chem* **381**, 921-35 (2000).

[10] Spellman, P.T. *et al.* Comprehensive identification of cell cycle-regulated genes of the yeast Saccharomyces cerevisiae by microarray hybridization. *Mol Biol Cell* **9**, 3273-97 (1998).

[11] Eisen, M.B., Spellman, P.T., Brown, P.O. & Botstein, D. Cluster analysis and display of genome-wide expression patterns. *Proc Natl Acad Sci U S A* **95**, 14863-8 (1998).

[12] Gilbert, D., Schroeder, M. & van Helden, J. Interactive visualization and exploration of relationships between biological objects. *Trends in Biotechnology* **18**, 487-495 (2000).

[13] van Helden, J., Andre, B. & Collado-Vides, J. Extracting regulatory sites from the upstream region of yeast genes by computational analysis of oligonucleotide frequencies. *J Mol Biol* **281**, 827-42 (1998).

[14] van Helden, J., Rios, A.F. & Collado-Vides, J. Discovering regulatory elements in non-coding sequences by analysis of spaced dyads. *Nucleic Acids Res* **28**, 1808-18 (2000).

[15] Brazma, A., Jonassen, I., Vilo, J. & Ukkonen, E. Predicting gene regulatory elements in silico on a genomic scale. *Genome Res* **8**, 1202-15 (1998).

[16] Graber, J.H., Cantor, C.R., Mohr, S.C. & Smith, T.F. Genomic detection of new yeast pre-mRNA 3'-end-processing signals. *Nucleic Acids Res* **27**, 888-94 (1999).

[17] Reinert, G. & Schbath, S. Compound Poisson and Poisson process approximations for occurrences of multiple words in Markov chains. *J Comput Biol* **5**, 223-53 (1998).

[18] van Helden, J., del Olmo, M. & Perez-Ortin, J.E. Statistical analysis of yeast genomic downstream sequences reveals putative polyadenylation signals. *Nucleic Acids Res* **28**, 1000-10 (2000).

[19] Karp, P.D. *et al.* The EcoCyc and MetaCyc databases. *Nucleic Acids Res* **28**, 56-59 (2000).

[20] Kanehisa, M. & Goto, S. KEGG: Kyoto Encyclopedia of Genes and Genomes. *Nucleic Acids Res* **28**, 27-30 (2000).

[21] Thomas, D. & Surdin-Kerjan, Y. Metabolism of sulfur amino acids in Saccharomyces cerevisiae. *Microbiol Mol Biol Rev* **61**, 503-32 (1997).

[22] Pellegrini, M., Marcotte, E.M., Thompson, M.J., Eisenberg, D. & Yeates, T.O. Assigning protein functions by comparative genome analysis: protein phylogenetic profiles. *Proc Natl Acad Sci U S A* **96**, 4285-8 (1999).

[23] Marcotte, E.M. *et al.* Detecting protein function and protein-protein interactions from genome sequences. *Science* **285**, 751-3 (1999).

[24] Marcotte, E.M., Pellegrini, M., Thompson, M.J., Yeates, T.O. & Eisenberg, D. A combined algorithm for genome-wide prediction of protein function [see comments]. *Nature* **402**, 83-6 (1999).

[25] Enright, A.J., Iliopoulos, I., Kyrpides, N.C. & Ouzounis, C.A. Protein interaction maps for complete genomes based on gene fusion events [see comments]. *Nature* **402**, 86-90 (1999).

[26] van Helden, J., Andre, B. & Collado-Vides, J. A web site for the computational analysis of yeast regulatory sequences. *Yeast* **16**, 177-87 (2000).

Author Index

Lecture Notes in Computer Science

For information about Vols. 1–1979
please contact your bookseller or Springer-Verlag